D0463508

Remedial Technologies for Leaking Underground Storage Tanks

Prepared by
ROY F. WESTON, INC.
UNIVERSITY OF MASSACHUSETTS

NOTICE

Remedial Technologies for Leaking Underground Storage Tanks

Prepared by

ROY F. WESTON, INC., Concord, California

L. M. Preslo, Project Manager; J. B. Robertson, Project Director; D. Dworkin, Project Engineer

UNIVERSITY OF MASSACHUSETTS, Environmental Science Program, Division of Public Health, Amherst, Massachusetts

E. J. Fleischer, Research Associate; P. T. Kostecki, Senior Research Associate; E. J. Calabrese, Professor

Originally Prepared for

Electric Power Research Institute, Palo Alto, California

M. E. McLearn and M. J. Miller, EPRI Project Managers, Heat, Waste, and Water Management Program, Coal Combustion Systems Division

and

USWAG/Edison Electric Institute, Washington, D.C.

W. Suyama, USWAG Project Manager, Utility Solid Waste Activities Group; W. A. Kucharski, Chairman, Underground Storage Tank Committee

Library of Congress Cataloging-in-Publication Data

Remedial technologies for leaking underground storage tanks.

Bibliography: p.
Includes index.
1. Petroleum products—Underground storage. 2. Oil storage tanks—Maintenance and repair. I. Electric Power Research Institute. II. Edison Electric Institute.
TP692.5.F46 1988 665.5′42 87-34251
ISBN 0-87371-125-4

Fourth Printing 1989

Third Printing 1988

Second Printing 1988

LEWIS PUBLISHERS, INC.
121 South Main Street, Chelsea, Michigan 48118

PRINTED IN THE UNITED STATES OF AMERICA

Preface

Tough new regulations proposed by the U.S. Environmental Protection Agency for underground storage tanks have alerted storage tank owners and managers to the pressing need for remedial action plans. Many states—including California, Maryland, Michigan, and New York—are going beyond the federal standards and implementing aggressive programs of their own.

The electric utility industry, which owns and operates many underground storage tanks for using, storing, and transferring petroleum products, has investigated, at length, various aspects of underground storage tank design and management technology.

This book presents the results of work undertaken by the Electric Power Research Institute (EPRI) and the Utility Solid Waste Activities Group (USWAG). An introduction to state-of-the-art cleanup technology is offered which includes a description and evaluation of available technologies for remediating soil and groundwater that contain petroleum products released from an underground storage tank leak, other discharges, or spills.

While not intended as a remedial design and implementation manual, the book does present information that will help industrial environmental managers and engineers, transportation fleet managers, fuel marketers and distributors, and insurance and risk managers, to plan with the new legal requirements in mind. A summary description and evaluation of 13 remedial methods for soil and groundwater cleanup are presented, organized in terms of four major considerations in evaluating the relative merit of each method: technical, environmental, economic, and implementation feasibility.

Underground storage tank owners and managers may become familiar with the methods which may be applicable to their particular situation, and may eliminate from further consideration those methods that are clearly unsuitable.

Contents

1 INTRODUCTION 1
Purpose and Scope 1
Format, Content, and Approach 2

2 SUMMARY OF REMEDIAL TECHNOLOGIES 5
Applicable Remedial Technologies 5
Environmental Fate of Hydrocarbon Constituents 7
Exposure Pathways 11
Relative Costs for Applicable Technologies 13
State Practices for Remedial Actions 15

3 IN SITU TECHNOLOGIES 17
Introduction 17
In Situ Volatilization (ISV) 17
In Situ Biodegradation 37
In Situ Leaching and Chemical Reaction 52
In Situ Vitrification 64
In Situ Passive Remediation 75
In Situ Isolation/Containment 84

4 NON-IN SITU TECHNOLOGIES 101
Introduction 101
Land Treatment Technology 101
Thermal Treatment 118
Asphalt Incorporation 135
Solidification/Stabilization 143
Groundwater Extraction and Treatment 157
Chemical Extraction 175
Excavation 183

5 EMERGING TECHNOLOGIES 195

REFERENCES 199

INDEX 211

List of Tables

2.1 Summary of Remedial Technologies 6
2.2 Common Constituents of Petroleum Products 8
2.3 Relative Environmental Partitioning of Petroleum Constituents Based on SESOIL Results 10
2.4 Categories of Migration Pathways 10
2.5 Exposure Pathways Associated with Various Steps of Remedial Technologies 12
2.6 Relative Cost Comparisons for Remedial Technologies 13
2.7 State Survey of Remedial Technology Practices 15
3.1 Factors That Influence the Distribution of Volatile Hydrocarbon Compounds in the Vadose Zone 20
3.2 Results of the Pilot Study for In Situ Volatilization of TCE 25
3.3 State Standards for Volatile Organic Compounds 34
3.4 Factors That Influence Biodegradation in Soils 42
3.5 Potential for Biodegradation of Selected Organic Pollutants in Unconfined Aquifers 50
3.6 Delivery Methods for In Situ Leaching 56
3.7 Matrix for Delivery Methods 57
3.8 Recovery Methods for In Situ Leaching................. 58
3.9 Surfactant Characteristics 58
3.10 In Situ Vitrification Studies 68
3.11 In Situ Vitrification Cost Estimate (1985 $) 76
3.12 Factors That Influence the Natural Processes for Passive Remediation 79
4.1 Factors Influencing Biodegradation of Hydrocarbon Compounds 107
4.2 Sludge Characteristics for Land Treatment— Case Study 2 110
4.3 Site Characteristics for Land Treatment— Case Study 2 110

4.4 Factors Monitored During Land Treatment—Case Study 2 111
4.5 Capital Costs for Land Treatment Operations 117
4.6 Operations and Maintenance Costs for Land Treatment Operations 118
4.7 Monitoring and Inspection Requirements 132
4.8 Advantages and Disadvantages of Cements and Pozzolans 150
4.9 Summary of Efficiency, Processing Time, and Relative Cost of Solidification/Stabilization Alternatives 156
4.10 Treatment Technologies for Groundwater Remedial Actions 165
4.11 Design Parameters for the Full-Scale Air Stripping Tower 167
4.12 Type of Petroleum Product and Applicable Technology for Groundwater Treatment 168
4.13 Hydrocarbon Removal Efficiency for Various Groundwater Treatment Technologies 170
4.14 Performance Characteristics of Landfill Equipment 188
4.15 Properties of Some Flammable Fuel Mixtures 190
5.1 Emerging Remedial Action Technologies 195

List of Figures

2.1 SESOIL: Schematic Presentation of the Soil Compartment (Cell) 9
3.1 Schematic of In Situ Stripping Pilot Study Apparatus 19
3.2 Major Factors Influencing In Situ Volatilization 21
3.3 Vent Pipe and Manifold System Layout 24
3.4 Service Station Site Plan 27
3.5 Schematic Plan for Soil Venting System with Gas Sampling Probes 28
3.6 Gasoline Recovery During Full-Scale In Situ Volatilization 29
3.7 Gasoline Recovery Rate—Monitor Well A During Full-Scale In Situ Volatilization 30
3.8 Schematic System for In Situ Biodegradation—Injection Well 39
3.9 Schematic System for In Situ Biodegradation—Infiltration Gallery 40
3.10 Schematic of a Leachate Recycling System 54
3.11 In Situ Vitrification Process Sequence 66
3.12 Engineering Scale In Situ Vitrification System 68
3.13 Large-Scale System Capabilities 70
3.14 Schematic for Large-Scale Off-Gas System 72
3.15 Typical Multilayer Cap System Profile 88
3.16 Keyed-In Slurry Wall/Hanging Slurry Wall 90
3.17 Upgradient Grout Curtain 91
3.18 Gilson Road Disposal Site 94
4.1 Mechanisms Which Occur During Landfarming 106
4.2 Air Temperature and Percentage of Residual Oil in Soil Following a Pipeline Break 109
4.3 Rotary Kiln Incinerator Schematic 120
4.4 An Example of a Circulating Bed Combustor 122

4.5 Schematic Illustration of the Low Temperature Thermal Stripping System 123
4.6 Typical Asphalt Batching Plant 136
4.7 Typical Lime—Fly Ash Pozzolanic Cement Plant 145
4.8 Envirosafe Stabilization Process 146
4.9 In Situ Surface Impoundment Treatment 147
4.10 Schematic of An Extraction/Injection Well System 158
4.11 Typical Air Stripping Installation for Groundwater Treatment 159
4.12 Oil/Water Separator 160
4.13 Typical Granular Activated Carbon Installation for Groundwater Treatment 162
4.14 Extended Aeration Treatment Plant for Above Ground Biodegradation 163
4.15 Product Recovery Using Water Table Depression 164
4.16 Air Stripping and Carbon Adsorption Treatment System Costs 174
4.17 Typical On-Site Solvent Extraction Process for Soils 177
4.18 Three Stages of Landfill Decomposition................. 185

SECTION 1

Introduction

PURPOSE AND SCOPE

The electric utility industry owns and operates many underground and aboveground storage tanks as well as other facilities for using, storing, or transferring petroleum products, primarily motor and heating fuels. The prevention, detection, and correction of leakage of these products from underground storage tanks (UST) has gained high priority in the utility industry and within the regulatory agencies. The 1984 amendments to the Resource Conservation and Recovery Act (RCRA) require the U.S. Environmental Protection Agency (EPA) to develop new Federal regulations for reducing and controlling environmental damage from underground storage tank leakage. In 1986, Congress authorized EPA to issue cleanup orders to owners and operators of leaking underground storage tanks and established a trust fund to finance cleanups in those cases where no owner or operator capable and willing to finance the cleanup is available. In addition, many states and localities have developed and are implementing strict regulations governing underground storage tanks and remedial actions for product releases to soil and groundwater.

Consequently, organizations such as the Electric Power Research Institute (EPRI) and the Utility Solid Waste Activities Group (USWAG) are conducting a series of technical investigations and evaluations of the various aspects of underground storage tank design and management technology, including tank design and installation, tank monitoring and inventory control, leak testing, tank removal and disposal, tank repair, remediation methods for petroleum hydrocarbons in soil and groundwater, fate and behavior of petroleum hydrocarbons in soil and groundwater, and economic analyses.

This report is the result of a cooperative effort between EPRI and USWAG's Leaking Underground Storage Tank Committee. It presents the results of one major component of the technical investigations — a description and evaluation of available technologies for remediating soil and groundwater that contain petroleum products released from an underground storage tank leak, other discharges, or spills.

It is the intent of this report to provide a general introduction to the state-of-the-art cleanup technology. The report includes a listing of feasible methods, a description of their basic elements, and some discussion of the factors to be considered in their selection and implementation for a remedial program. *This report is not intended to be a remediation design and implementation manual.* All soil and groundwater remediation problems require site-specific considerations in choosing, designing, and implementing the most appropriate remedial programs.

The report provides sufficient information to enable the readers to become familiar with the methods which may be applicable to their specific problems and eliminate from further consideration those methods which are clearly inappropriate. Neither USWAG, EPRI, the Edison Electric Institute (EEI), nor any of their members endorses or recommends any particular cleanup method as it relates to a particular remediation problem.

FORMAT, CONTENT, AND APPROACH

This report is a summary description and evaluation of 13 remedial methods for soil and groundwater cleanup. There are many technical, economic, institutional, and related factors to be considered in designing or evaluating remedial techniques. Many of these factors are highly site-specific; therefore, there is no single method which is best for all or even most circumstances.

Selecting and implementing a remedial method depends first of all on a determination of cleanup goals, which are the contaminant levels or concentration limits to which the site must be cleaned. These, in turn, are usually based on either an assessment of potential risk or regulatory standards. Remedial options are then chosen by assessing the feasibility of each option to achieve the desired cleanup goal and evaluating the relative cost and acceptability of the method. The method selected may not always be the most cost effective.

The information in this report is organized and presented in terms of four major considerations in evaluating the relative feasibility of each method:

- Technical feasibility.
- Implementation feasibility.
- Environmental feasibility.
- Economic feasibility.

Technical Feasibility

The technical feasibility discussion of each method includes the following components:

- General description of the basic technical principles involved with the method.
- Brief description of one or more typical case histories where the method has been applied.
- Implementation considerations such as:
  - Design and operation.
  - Equipment requirements.
  - Disposal issues.
  - Monitoring requirements.
  - Permitting requirements.

Implementation Feasibility

The feasibility of implementing each method is discussed in terms of practical considerations that will influence the applicability of a technology to a particular site. Selection criteria for implementation feasibility include:

- Design considerations.
- Equipment requirements.
- Treatment needs.
- Disposal needs.
- Monitoring requirements.
- Permitting requirements.
- In the case of non-in situ technologies, on-site versus off-site operation.

Environmental Feasibility

The environmental feasibility of each method is based primarily on the potential effectiveness of the method to achieve environmental goals for all media of concern—soil, groundwater, surface water, and air—as well as exposures to biota in each of those media. Environmental feasibility is based on risk analysis methods. A risk analysis involves two interdependent factors:

1. Pathways through which the substances can reach biological receptors, such as fish or humans.

2. Exposures which each biological receptor might receive from the various migration pathways.

Risk analysis can then be performed by combining pathway analyses with potential exposure risk and assessing the resultant health or damage risk. Risk analysis must, by necessity, be based on site-specific considerations. Therefore, this report presents qualitative and generic information on the types of exposure pathways most commonly associated with each method and factors that must be considered in assessing potential exposures.

Economic Feasibility

The economic feasibility of any method should be based on a comparison of cost effectiveness and cleanup effectiveness (degree of reduction in total risk). This, again, can only be done by taking into consideration many site-specific factors. Nonetheless, some general cost comparisons can be made among the various methods if certain assumptions are made. The comparisons presented in this report are useful for understanding relative cost differences under certain circumstances and for recognizing important factors that may greatly influence relative costs. Economic information presented in this report should not be used as a basis for selecting a remedial option, but rather for guiding the planning, screening, and selection process.

Cost information has been converted to 1986 U.S. dollars and is provided for the following aspects of each method:

- Capital costs.
- Installation costs.
- Operation and maintenance costs.
- Relative cost effectiveness.

References

Additional information on the methods discussed is available from numerous references and other sources. Key references for each technology are cited in the text and listed at the end of this report.

SECTION **2**

Summary of Remedial Technologies

This document provides information on the technical, implementation, environmental, and economic feasibility of 13 remedial technologies for soils and/or groundwaters impacted by leaking underground storage tanks (UST). The remedial technologies described have been divided into two categories: in situ treatment and non-in situ treatment. In situ treatment refers to treatment of soil or groundwater in place. The remedial technologies have been further subdivided within these categories as follows:

- In Situ Technologies:
  - Volatilization
  - Biodegradation
  - Leaching and Chemical Reaction
  - Vitrification
  - Passive Remediation
  - Isolation/Containment
- Non-In Situ Technologies:
  - Land Treatment
  - Thermal Treatment
  - Asphalt Incorporation
  - Solidification/Stabilization
  - Groundwater Extraction and Treatment
  - Chemical Extraction
  - Excavation

APPLICABLE REMEDIAL TECHNOLOGIES

Table 2.1 summarizes information for each remedial technology with regard to potential exposure pathways, application of the technology to specific petroleum products, advantages and limitations of each technology, and costs. Addi-

Table 2.1 Summary of Remedial Technologies

Technology	Exposure Pathways[1]	Applicable Petroleum Products[2]	Advantages	Limitations	Relative Costs[3]
In Situ					
Volatilization	1-7	1, 2, 4	Can remove some compounds resistant to biodegradation.	VOCs only.	Low
Biodegradation	1-7	1, 2, 4	Effective on some nonvolatile compounds.	Long-term timeframe.	Moderate
Leaching	1-7	1, 2, 4	Could be applicable to wide variety of compounds.	Not commonly practiced.	Moderate
Vitrification	1-7	1, 2, 3, 4		Developing technology.	High
Passive	1-7	1, 2, 3, 4	Lowest cost and simplest to implement.	Varying degrees of removal.	Low
Isolation/ Containment	1-7	1, 2, 3, 4	Physically prevents or impedes migration.	Compounds not destroyed.	Low to moderate
Non-In Situ					
Land treatment	1-7	1, 2, 3	Uses natural degradation processes.	Some residuals remain.	Moderate
Thermal Treatment	1-6	1, 2, 3, 4	Complete destruction possible.	Usually requires special facilities.	High
Asphalt Incorporation	1-6	1, 2	Use of existing facilities.	Incomplete removal of heavier compounds.	Moderate
Solidification	1-6	1, 2, 3, 4	Immobilizes compounds.	Not commonly practiced for soils.	Moderate

Table 2.1, continued

Technology	Exposure Pathways[1]	Applicable Petroleum Products[2]	Advantages	Limitations	Relative Costs[3]
Groundwater Extraction and Treatment	1-6	1, 2, 4	Product recovery, groundwater restoration.		Moderate
Chemical Extraction	1-6	1, 2, 3, 4		Not commonly practiced.	High
Excavation	1-6	1, 2, 3, 4	Removal of soils from site.	Long-term liability.	Moderate

[1]Exposure pathways: 1 = Vapor Inhalation; 2 = Dust Inhalation; 3 = Soil Ingestion; 4 = Skin Contact; 5 = Groundwater; 6 = Surface Water; and 7 = Plant Uptake.

For additional information regarding primary versus secondary exposure pathways, consult the subsection on exposure pathways for a particular technology, or see Table 2.5.

[2]Applicable petroleum products: 1 = gasolines; 2 = fuel oils (#2, diesel, kerosenes); 3 = coal tar residues; and 4 = chlorinated solvents.

For further information regarding the applicability of a specific technology to specific petroleum types, refer to the subsection on environmental effectiveness for that technology.

[3]Costs are highly dependent on site conditions. For additional information on costs, refer to Table 2.6, or consult the subsection on economic feasibility for a specific technology.

tional information on exposure pathways is presented in Tables 2.6 and 2.7. Detailed discussions of potential exposure pathways and cost are presented in Section 3, In Situ Technologies, and Section 4, Non-In Situ Technologies.

ENVIRONMENTAL FATE OF HYDROCARBON CONSTITUENTS

The problems caused by hydrocarbon spillage or disposal are complex for scientists, regulatory officials and industrial managers. Risk assessment is a management tool that can be used as an aid in the evaluation of the public health risks associated with chemical releases to soil and groundwater. The risk assessment process involves hazard identification (the examination of human and animal toxicity data for a particular chemical) and exposure assessment. Exposure assessment involves the evaluation of the potential routes of exposure following a chemical release, as well as the possible degree of human exposure from each of the identified exposure pathways. A comprehensive risk analysis involves a synthesis of both hazard and exposure assessment information. The risk of exposure through skin contact or ingestion is related to the tendency of a particular compound to adsorb onto soil particles and also into vegetation through root uptake which may enter the food chain. Likewise, the risk of exposure through inhalation is related to the tendency of a compound to volatilize either directly from soil or from water sources that have become impacted subsequent to releases of petroleum products. Similarly, the solubility and the specific gravity of a compound influence the way and the degree to which a chemical can impact surface or groundwater supplies, thereby controlling the risk of exposure through these pathways.

Petroleum products (gasoline, fuel oil, etc.) are complex mixtures of hydrocarbons. From 100 to 150 compounds can be identified in a typical gasoline, although many more are known to be present. Each constituent in a mixture has different physical and chemical characteristics that control the behavior of petroleum products in a soil system. Alternatively, insights that may apply to the study of complex fuel mixtures can be gained through the study of a number of individual hydrocarbon compounds. Research efforts by Fleischer et al. (1986) have centered on 13 specific compounds because of their use in petroleum products, their tendency to be released to the subsurface environment, and their potential toxicity. These compounds are listed in Table 2.2.

The potential environmental fate of these organic compounds was examined by conducting computer simulations using the unsaturated zone environmental fate model SESOIL. The Seasonal Soil Compartment Model (SESOIL) was developed by Bonazountas and Wagner at A.D. Little, Inc. for the U.S. EPA Office of Toxic Substances (1984). The theoretical soil column approach utilized by SESOIL is schematically shown in Figure 2.1. SESOIL has been used by Bonazountas and Wagner to evaluate the potential environmental effects of two land treatment sites (1981) and the potential fate of six buried solvents (1983). A testing program conducted by Anderson-Nichols, Inc. for the EPA

Table 2.2 Common Constituents of Petroleum Products

Gasoline and Fuel Oils	Heavy Oils and Waste Oils
Benzene	Benz(a)Anthracene
Ethylbenzene	Benzo(a)Pyrene
(n) Heptane	Naphthalene
Pentane	Phenanthrene
(n) Hexane	
1-Pentene	
(o) Xylene	
Toluene	
Phenol	

Source: Fleischer et al., 1986.

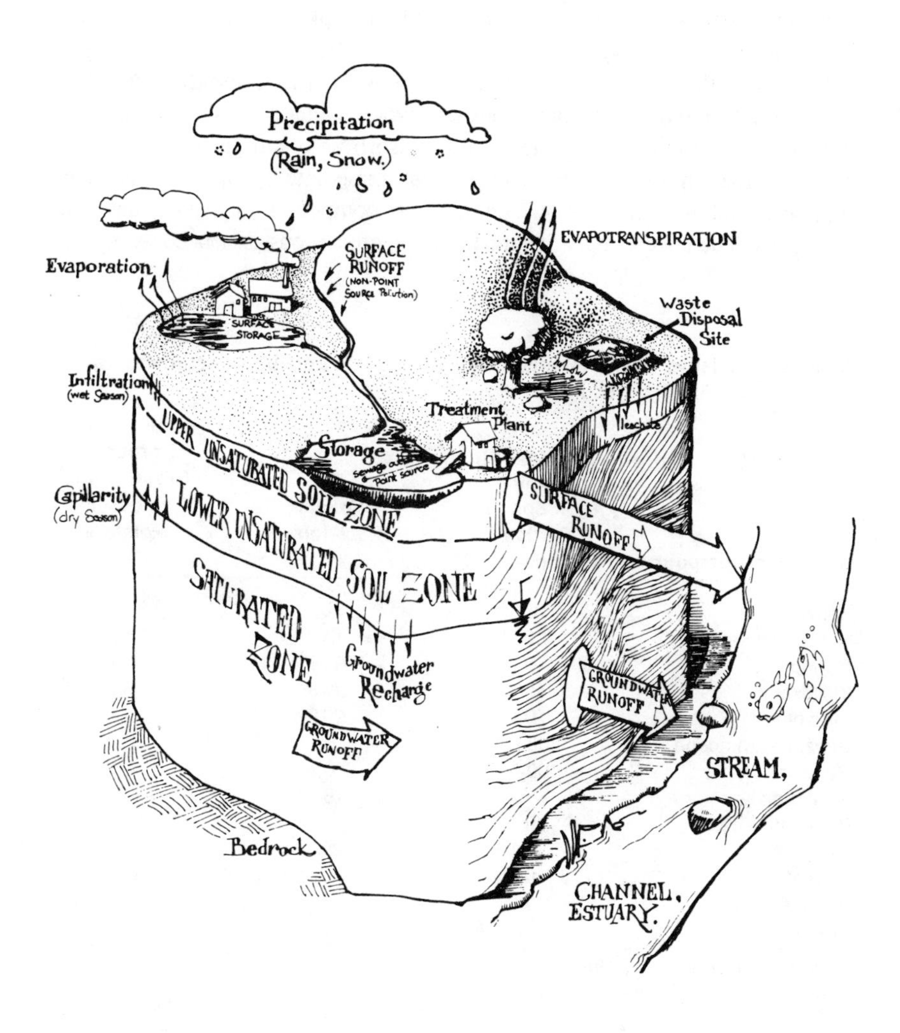

Reprinted from
Proceedings of the 1984 Speciality Conference on Environmental Engineering
ASCE/Los Angeles, CA, June 25-27, 1984

Figure 2.1 SESOIL: Schematic presentation of the soil compartment (cell).

concluded that the SESOIL model could be a useful model for chemical mobility and fate screening (Watson and Brown, 1984).

Results of the SESOIL simulations are presented in Table 2.3. Based on these results, the organic compounds that were evaluated can be divided into four groups: 1) those that preferentially adsorb onto soil particles; 2) those that volatilize rapidly; 3) those that pose an immediate threat to groundwater supplies; and 4) those for which no one compound migration pathway dominates. The results imply that the lighter hydrocarbons associated with gaso-

Table 2.3 Relative Environmental Partitioning of Petroleum Constituents Based on SESOIL Results

Petroleum Compound	Adsorption Onto Soil Particles (%)	Volatilization (%)	Soluble Portion in Groundwater and Soil Moisture (%)
Benzene	3	62	35
Ethylbenzene	21	59	20
(n)Heptane	0.1	99.8	0.1
(n)Hexane	0.1	99.8	0.1
(n)Pentane	0.1	99.8	0.1
Benz(a)Anthracene	100	0	0
Benzo(a)Pyrene	100	0	0
Naphthalene	61	8	31
Phenanthrene	88	2	10
1-Pentene	0.1	99.8	0.1
Phenol	9	0.01	91
Toluene	3	77	20
(o)Xylene	15	54	31

Source: Fleischer et al., 1986

Table 2.4 Categories of Migration Pathways

Adsorb to Soil Particles	Volatilize in Air	Solubilize in Groundwater	Multiple Pathways
Benzo(a)Pyrene	(n)Hexane	Phenol	Benzene
Phenanthrene	(n)Heptane		Ethylbenzene
Benz(a)Anthracene	(n)Pentane		Napthalene
	1-Pentene		Toluene
			(o)Xylene

Source: Fleischer et al., 1986

line are more likely to volatilize whereas heavier constituents can be bound tightly to soil particles. Table 2.4 presents these compound groupings.

These results have implications regarding both human exposure pathways and the potential effectiveness of various remedial actions. For those compounds that tend to volatilize rapidly, the primary exposure pathway can be expected to be vapor inhalation. An effective remedial action for soils containing these compounds might take advantage of their volatile nature. In situ volatilization is one such process. For compounds that adsorb tightly to soil, skin contact or soil ingestion are important exposure pathways. In situ volatilization would not be an effective remedial action in such a case; however, the use of a soil cover system might provide a solution to the problem. Multiple exposure and migration pathways are possible for many of the compounds contained in petroleum products. Thus, the minimization of health risks associated with exposure to soils containing petroleum products as well as the maximization of benefits obtained from remedial action efforts requires serious consideration.

EXPOSURE PATHWAYS

Table 2.5 summarizes information on potential exposure pathways resulting from each of the 13 remedial options. Exposure pathways can be broken down into two categories: 1) direct human exposure pathways; and 2) environmental exposure pathways. These two categories can be further subdivided into primary exposure pathways and secondary exposure pathways.

Primary pathways are those which 1) directly affect site operations and personnel (e.g., vapor inhalation or skin contact during groundwater sampling); or 2) directly affect cleanup levels which must be achieved by the remedial technology. For example, when groundwater impact is the principal issue at a site, groundwater impact sets the cleanup level and corresponding timeframe when cleanup ceases.

Secondary pathways are less important than primary pathways and occur as a minor component during site operations (e.g., particle ingestion during monitoring as a result of soil sampling or wind blown dust). They also exhibit significant decreases with time as treatment progresses.

Table 2.5 also details exposure pathways that may be encountered during various aspects of the remediation process. The aspects include excavation, installation, operations and maintenance, and monitoring.

A quantitative evaluation (or risk analysis) of exposure pathways and exposure levels that result from remedial action at a particular site would require detailed monitoring and study. The exposure discussions presented throughout this document indicate only possible exposure through a particular

Table 2.5 Exposure Pathways Associated with Various Steps of Remedial Technologies[1]

Remedial Technology	Various Steps of Remedial Technologies													
	Vapor Inhalation		Dust Inhalation		Particle Ingestion		Skin Contact		Groundwater		Surface Water		Plant Uptake	
	EXPOSURE PATHWAYS													
In Situ	*1*	*2*	*1*	*2*	*1*	*2*	*1*	*2*	*1*	*2*	*1*	*2*	*1*	*2*
Volatilization	IOM	—	I	OM	I	OM	IOM	—	IOM	—	IOM	—	—	IOM
Biodegradation	IOM	—	I	OM	I	OM	IOM	—	IOM	—	IOM	—	—	IOM
Leaching	IOM	—	I	OM	I	OM	IOM	—	IOM	—	IOM	—	—	IOM
Vitrification	IOM	—	I	OM	I	OM	IOM	—	IOM	—	IOM	—	—	IOM
Passive	OM	—	—	OM	—	OM	M	O	OM	—	OM	—	—	OM
Isolation/ Containment	IOM	—	I	OM	I	OM	IOM	—	IOM	—	IOM	—	—	IOM
Non-In Situ														
Land Treatment	EIOM	—	EIO	M	EIO	M	EIOM	—	EIOM	—	EIOM	—	—	IOM
Thermal Treatment	EIOM	—	EIO	—	EIO	—	EIO	—	E	—	E	—	—	—
Asphalt Incorporation	EOM	—	EO	—	EO	—	EO	—	EO	—	EO	—	—	—
Solidification	EIOM	—	EIO	—	EIO	—	EIO	—	EIO	—	EIO	—	—	—
Pump and Treat	EIOM	—	EI	OM	EI	OM	EIOM	—	EIOM	—	EIOM	—	—	—
Chemical Extraction	EIOM	—	EI	OM	EI	OM	EIOM	—	EIOM	—	EIOM	—	—	—
Excavation	E	—	E	—	E	—	E	—	E	—	E	—	—	—

[1]Notes explaining the abbreviations in this table are presented below.

Notes:

1. The steps of each remedial technology are abbreviated as follows:
   E=Excavation
   I=Installation
   O=Operations and Maintenance
   M=Monitoring (refers to monitoring site and/or treatment emissions, if applicable).
2. The exposure pathways are classified as:
   1=Primary
   2=Secondary
   The exposure pathways and their justification for being classified as primary (1) or secondary (2) are discussed in detail in the subsections on exposure pathways for each remedial technology (Sections 3 and 4).

Table 2.5, continued

Primary pathways are those 1) which directly affect site operations and personnel (e.g., vapor inhalation or skin contact during groundwater sampling); or 2) cleanup levels which must be achieved by the remedial technology. For example, when groundwater impacts are an issue at a site, groundwater impacts set the cleanup levels and corresponding timeframe when cleanup ceases.

Secondary pathways are those which are less important than primary pathways, occur as a minor component during site operations (e.g., particle ingestion during monitoring as a result of soil sampling or wind blown dust), or show significant decreases with time as treatment progresses.

pathway and in no way imply the extent of exposure or any particular health risks associated with that exposure. A quantitative site-specific risk analysis must be performed to accomplish the latter task.

RELATIVE COSTS FOR APPLICABLE TECHNOLOGIES

It is important to note that site-specific conditions will dictate the actual design decisions which, in turn, will dramatically affect the final cost for each technology. In order to develop relative cost comparisons for these technologies, cost ranges were derived for each technology by assuming a hypothetical site of moderate size. The size of this site, as well as the relative costs and the hypothetical quantities of impacted soil and groundwater, are presented in Table 2.6. Key design decisions such as off-site versus on-site operation and leased purchased equipment were also specified.

Table 2.6 Relative Cost Comparisons for Remedial Technologies

Remedial Technologies	Relative Total Cost[1]	Design Assumptions[2]
In Situ		
Volatilization	Low	7.63 meters (25 ft) centers for 8 venting pipes; no treatment for effluent gases.
Biodegradation	Moderate	3 extraction wells with infiltration galleries for injection; flow rate of 0.002572 m^3/sec (40 gpm) through reactor.
Leaching	Moderate	Same assumptions as biodegradation except treatment unit differs.

Table 2.6, continued

Remedial Technologies	Relative Total Cost[1]	Design Assumptions[2]
Vitrification	High	Unit costs are based on a larger site (Pacific Northwest Laboratories, 1986).
Passive Remediation	Low	Monitoring costs only; 4 monitoring wells with quarterly sampling of aromatic volatile hydrocarbon indicator compounds.
Isolation/ Containment	Low	Cap composed of liner, soil, and bentonite; no slurry wall.
	Moderate	Same cap with slurry wall.
Non-In Situ		
Land Treatment	Moderate	Purchased (not leased) equipment; on-site operation.
Thermal Treatment	High	Leased mobile unit; on-site operation.
Asphalt Incorporation	Moderate	Off-site operation; shipping costs are additional.
Solidification	Moderate	Leased equipment; 30 percent portland cement, 2 percent sodium silicate.
Groundwater Treatment	Low to Moderate	Moderately-sized carbon unit or air stripper without effluent treatment.
Chemical Extraction	High	Leased mobile unit; on-site operation.
Excavation and Landfill	Moderate to High	Leased equipment; costs relative to landfill disposal fees and transportation costs.

[1]Unit costs:
Low = Less than $13.00/m^3 ($10/yd^3) of soil or 3,780 liters (1,000 gallons) of water.
Moderate = 13 to $130/m^3 ($10-$100/yd^3) of soil or 3,780 liters (1,000 gallons) of water.
High = Greater than $130/m^3 ($100/yd^3) of soil or 3,780 liters (1,000 gallons) of water.

[2]Site with dimensions of 30.5m×15.25m×6.1m (100 ft×50 ft×20 ft) depth, a volume of 2,837.3m^3 (3,700 yd^3) weighing 3,636.4 metric tons (4,000 tons) and 37,800,000 liters (10,000,000 gallons) of impacted groundwater. Depth to the water table is 6.1 meter (20 ft).

STATE PRACTICES FOR REMEDIAL ACTIONS

Table 2.7 provides a state-by-state summary of remedial technology practices that have been permitted or used (Kostecki et al., 1985). Information for Table 2.7 was acquired through a national survey using both mail and follow-up telephone surveys.

The issues covered in these surveys were:

- Type of remedial options that are currently allowed.
- State regulations that trigger site investigations and existing levels for cleanup.
- Analytical techniques used for sample analysis.
- Human health effects.

The results of various remedial options that have been used are summarized in Table 2.7; other results are detailed in Kostecki et al. (1985).

It should be noted that data in the table are incomplete and in some cases out of date. Nonetheless, the results are useful in demonstrating in a qualitative sense which remedial technologies have been most popularly accepted and most widely applied in recent years. The most commonly used methods are land treatment, air stripping, and land filling.

Table 2.7 State Survey of Remedial Technology Practices

State	Land Treatment	Air Stripping of Groundwater	Approved State	Hazardous Waste	As Cover Material	Road Use	Aeration of Soil	In Situ Biodegradation	Asphalt Batching	Land Spreading	General Road Use	Leave in Place	Moderate	Soil Venting	Cement Kiln	Heat in Rotary Kiln	Soil Shredding
Alabama	X		X	X													
Alaska	X							X									
Arizona		X		X			X										
Arkansas	X	X	X					X									
California	X		X	X			X										
Colorado			X														
Connecticut	X	X															
Delaware	X		X	X		X										X	
Florida			X	X			X	X					X			X	
Georgia					X	X											
Idaho		P										X					
Illinois			X														
Indiana	X		X				X										
Iowa	X	X															

Table 2.7, continued

State	Land Treatment	Air Stripping of Groundwater	Approved State	Hazardous Waste	As Cover Material	Road Use	Aeration of Soil	In Situ Biodegradation	Asphalt Batching	Land Spreading	General Road Use	Leave in Place	Moderate	Soil Venting	Cement Kiln	Heat in Rotary Kiln	Soil Shredding
Kansas	X	X			X												
Kentucky	X		X	X													
Louisiana	X	X	X	X	X					X			X				
Maine	P	X	X														
Maryland	X		X										X				
Massachusetts		X	X	X				X	X								
Michigan	X		X														
Minnesota	X	X		X	X			X	X	X							
Mississippi	X		X	X													
Missouri	X		X														
Montana			X	X										X	X		
Nebraska	X	X			X						X						
Nevada		X	X								X						
New Hampshire		X	X						X								
New Jersey	X	X	X														
New Mexico	X																
New York	O	X	X	X													
North Carolina			X														
North Dakota			X								X						
Ohio			X	X				X		X	X						
Oklahoma			X	X													
Oregon	X		X														
Pennsylvania	X	X	X		X												
Rhode Island		X							X								
South Carolina		X	X														
South Dakota	X	X					X										
Tennessee	X											X					
Texas	X		X														
Utah				X								X					
Vermont				X					X								
Virginia	X		X														
Washington	X	X						X									
West Virginia	X			X													
Wisconsin			X		X		X							X			
Wyoming	X				X												
Total	29	19	31	17	8	1	7	7	5	3	4	3	3	2	1	1	1

Source: Kostecki et al., 1985

P = Proposed, not tabulated in totals O = Attempted one time

Note: This table presents the remedial technologies that have been accepted and implemented within each state.

SECTION 3

In Situ Technologies

INTRODUCTION

The remedial technologies discussed in this section address only those technologies that can be performed in situ, or in place, at the site. No excavation is required, so exposure pathways are minimized. The only exposure pathways which must be considered are those that result from the actual streams produced by the in situ technologies and not those associated with the handling and transport involved in the non-in situ technologies discussed in Section 4.

The in situ technologies discussed in this section include: In Situ Volatilization, Biodegradation, Leaching and Chemical Reaction, Vitrification, Passive Remediation, and Isolation and Containment.

IN SITU VOLATILIZATION (ISV)

General Description

In situ volatilization (ISV) is the process by which volatile compounds are removed from the in-place soil through the utilization of forced or drawn air currents. Depending upon the types of compounds that are present and site conditions, in situ volatilization can be a very effective, cost-efficient remedial action. ISV has been successfully performed at many sites. This subsection describes in detail the feasibility of this process.

Process Description

In situ volatilization or in situ air stripping involves the removal of volatile organic compounds from subsurface soils by mechanically drawing or venting air through the soil matrix. A variation of this process involves the positive pressure flow of air from a potentially impacted site (e.g., a building basement) to divert petroleum hydrocarbon flow and therefore mitigate impact.

Figure 3.1 depicts an example of a system used to enhance subsurface ventilation and volatilization of volatile organic compounds. The unit operations represented are common to most in situ volatilization systems, although such systems vary considerably in size and design because of site-specific requirements. The basic operations are:

- A pre-injection air heater warms the influent air to raise subsurface temperatures and increase the volatilization rate. Often the soil acts as a vast heat sink and subsurface temperature rises are not appreciable. In cold climates, however, air heaters are valuable for system freeze protection.
- Injection and/or induced draft fans establish the air flow through the unsaturated zone.
- Slotted or screened pipe is often used to allow air to flow through the system but restrict entrainment of soil particles.
- A treatment unit, often activated carbon, is used to recover volatilized hydrocarbons thereby minimizing air emissions. The effluent from this unit must comply with air pollution standards discussed later in this subsection.
- Miscellaneous air flow meters, bypass and flow control valves, and sampling ports are generally incorporated into the design to facilitate air flow balancing and assess system efficiency.

Technical Feasibility of In Situ Volatilization

Technical Description

Petroleum products are complex mixtures that contain hundreds of distinct hydrocarbons. The varying composition of these products results in differing chemical and physical properties which make some products more amenable than others to removal by in situ volatilization. Generally, compounds with higher vapor pressure and lower solubility in water are more efficiently removed or stripped. The Henry's Law Constant, a measure of the equilibrium distribution between air and dilute solutions, is typically used as an indicator of stripability. Because of this property and the higher content of volatile organic compounds in gasoline, proportionally more gasoline will be removed by in situ volatilization than kerosene or heavy heating oils.

Hydrocarbon compounds in the unsaturated zone can volatilize and diffuse both to groundwater and to the surface as well as be directly transported to

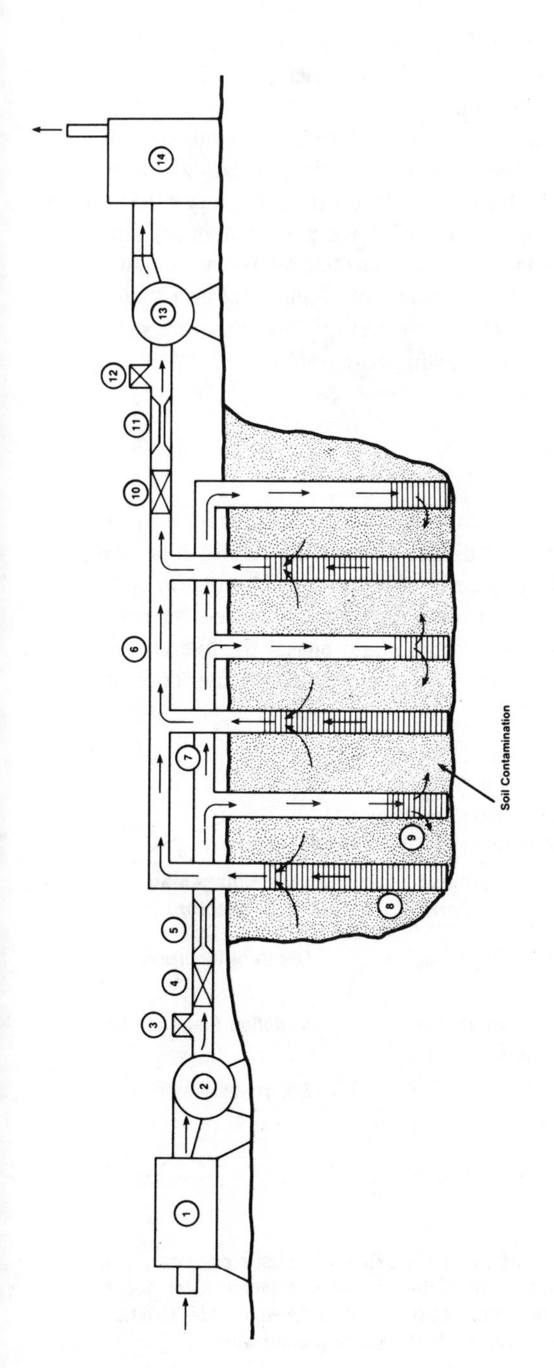

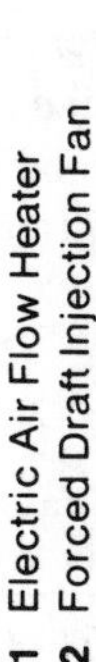

1 Electric Air Flow Heater
2 Forced Draft Injection Fan
3 Injection Air Bypass Valve
4 Injection Air Sampling Port
5 Injection Air Flow Meter
6 Extraction Manifold
7 Injection Manifold
8 Slotted Vertical Extraction Vent Pipe (typ)
9 Slotted Vertical Injection Vent Pipe (typ)
10 Extraction Air Sampling Port
11 Extraction Air Flow Meter
12 Extraction Air Bypass Valve
13 Induced Draft Extraction Fan
14 Vapor Carbon Package Treatment Unit

Figure 3.1 Schematic of in situ stripping pilot study apparatus.

the groundwater. Possible transport mechanisms include 1) liquid volatilization to vapor and diffusion to the surface; and 2) volatilization from the surface of particles and subsequent diffusion.

The controlling mechanism for vapor and chemical diffusion is site-specific and may not be easily evaluated. The factors influencing volatilization of hydrocarbon compounds from soils have been documented by Jury (1986). These factors fall into four categories: Soil, Environment, Chemical, and Management. Table 3.1 lists these factors and notes their relative importance to evaluating the applicability of in situ volatilization. Figure 3.2 graphically presents the influence of two major factors, permeability and chemical volatility. Each factor is discussed in the following paragraphs.

• *Soil Factors*

Water content—Water content influences the rate of volatilization by affecting the rates at which chemicals can diffuse through the vadose zone. An increase in soil water content will decrease the rate at which volatile compounds are transported to the surface via vapor diffusion. However, if a compound has an appreciable concentration in the liquid phase, the overall rate of transport to the surface may not be dramatically affected by water content.

Table 3.1 Factors That Influence the Distribution of Volatile Hydrocarbon Compounds in the Vadose Zone

Soil Factors	Environmental Factors	Chemical Factors	Management Factors
Water content	Temperature properties	Chemical	Depth of incorporation
Porosity/ Permeability	Wind	Chemical Quantities	Irrigation management
Clay content	Evaporation		Soil management
Adsorption site density	Precipitation		

Source: Jury, 1986

Notes:

1. Soil factors and chemical properties influence the extent to which a compound will diffuse or volatilize within the vadose zone more than do the other factors listed above.
2. Precipitation and irrigation management must also be considered in order to calculate the potential for downward leaching of chemicals to groundwater.

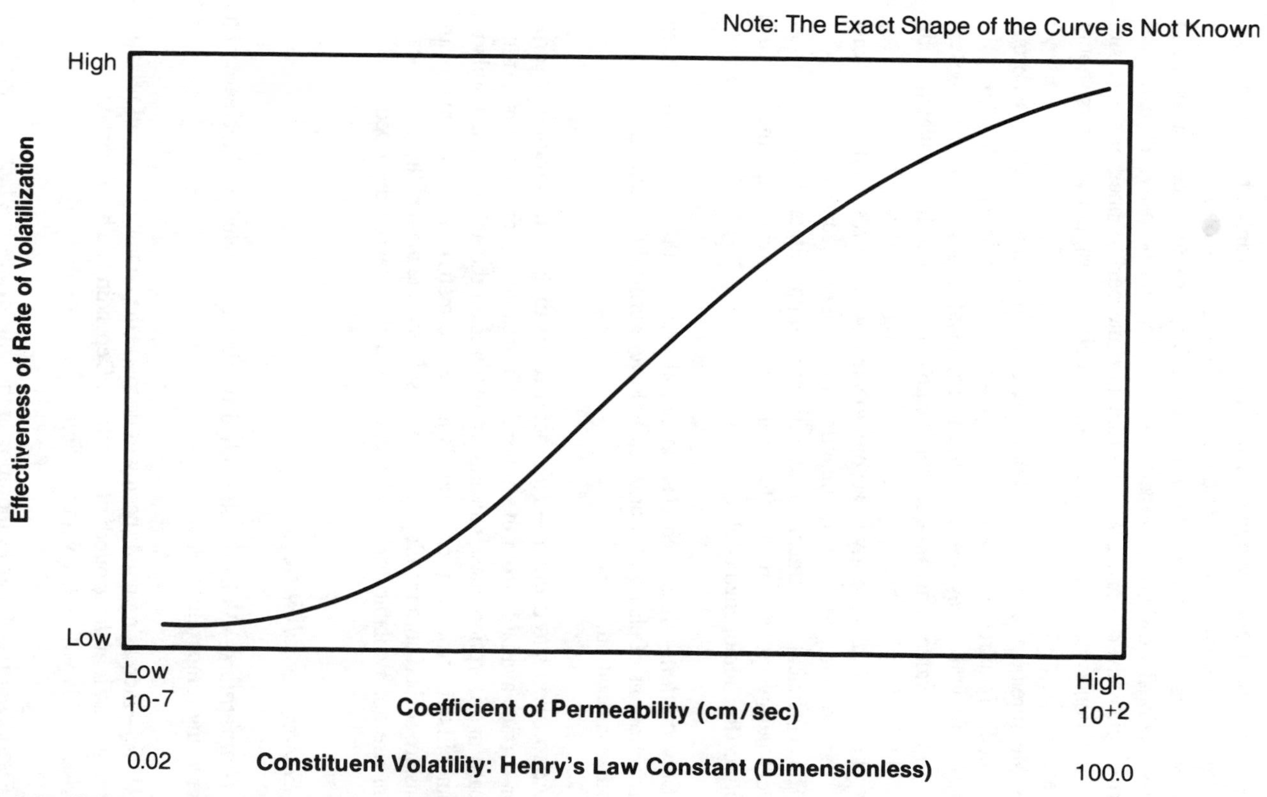

Figure 3.2 Major factors influencing in situ volatilization.

Porosity and permeability—The rate at which hydrocarbon compounds volatilize and are transported to the surface is a function of the travel distance and cross-sectional area available for flow. Diffusion distance increases and cross-sectional flow area decreases with decreasing porosity. Porosity and permeability are of particular importance to the performance of the in situ volatilization system, and the following associated design parameters should be evaluated:

- Measurement of air flow and pressure drop through the subsurface porous media requires installation of well screen and field testing.
- Air flow and pressure drop due to frictional losses through the system piping, valves, instruments and treatment units can be calculated using fluid flow principles.
- Blower or fan capacity to accommodate flow and pressure requirements can be obtained from the manufacturer's blower curves.

The coefficient of permeability is a function of porosity and can range from a very permeable 10^2 cm/sec (for clean gravel) to a very impermeable 10^{-7} cm/sec (for homogeneous clay).

Clay content—Increased clay content decreases soil permeability. High concentrations of clay will dramatically lower permeability and therefore inhibit volatilization.

Adsorption site density—This factor refers to the concentration of sorption surface available from the mineral and organic contents of soils. An increase in adsorption sites indicates an increase in the ability of the soils to immobilize hydrocarbon compounds in the soil matrix. When the hydrocarbons are sorbed onto the clay surfaces or soil organic matter, these hydrocarbons are not available for volatilization until they can be desorbed.

- *Environmental Factors*

Temperature—Volatilization of hydrocarbon compounds will increase with increasing temperature.

Wind—Increasing wind speed will decrease the boundary layer of relatively stagnant air at the ground/air interface. Depending on hydrocarbon characteristics, this will assist volatilization.

Evaporation—Water evaporation at the soil surface is a factor controlling the upward flow of water through the unsaturated zone. To the extent that this flow also transports hydrocarbon compounds dissolved in water to the soil surface, volatilization is enhanced by evaporation.

Precipitation—Precipitation provides water for infiltration into the vadose zone. Hydrocarbon compounds which are more prone to leaching and less prone to soil adsorption will be influenced.

- *Chemical Factors*

Chemical factors are critical mechanisms directly affecting the way in which various hydrocarbon compounds interact with the soil matrix. Chemical properties such as solubility, concentration, octanol-water partitioning coefficient, and vapor pressure affect the susceptibility of chemicals to the in situ volatilization process. The measure of volatilization used most often is the Henry's Law Constant. The more volatile compounds such as benzene will have a dimensionless constant of approximately 0.25 or greater.

The less volatile compounds, such as napthalene, will have a constant closer to 0.02. However, the in situ volatilization process is complex, and its success cannot be predicted on the basis of this or any single factor.

- *Management Factors*

The management factors that alter soil conditions can influence volatilization. The influence of such techniques (i.e., discing, fertilization, irrigation, etc.) has more applicability to land-farming procedures which are discussed in detail in Section 4. Generally, soil management techniques which decrease leaching, increase soil surface contaminant concentrations, or maximize soil aeration will assist volatilization.

Experience

In situ volatilization has been increasingly used during the past few years, and selected case studies are presented here. Although presented for specific chemical compounds, the techniques can be used for most volatile compounds.

Pilot-Scale. Coia, Corbin, et al. (1985) demonstrated two pilot systems during a 14-week field program at a site comprised of sandy glacial soils where open burning of waste solvents had been performed periodically during the last 30 years. Trichloroethylene (TCE) was found in the groundwater at levels two to three orders of magnitude greater than the recommended drinking water quality criterion of 2.7 μg/L (ppb), and soil concentrations were found as high as 5,000 to 7,000 mg/kg (ppm).

The experimental apparatus for the in situ pilot demonstration was comprised of a pipe vent and forced air ventilation system along with the necessary control mechanisms to monitor and sample the volatilization process. The general operation shown in Figure 3.3 entailed continuous injection and extraction of air through a series of pipe vents installed in the unsaturated zone in addition to monitoring, sampling, and analysis of the exhaust air.

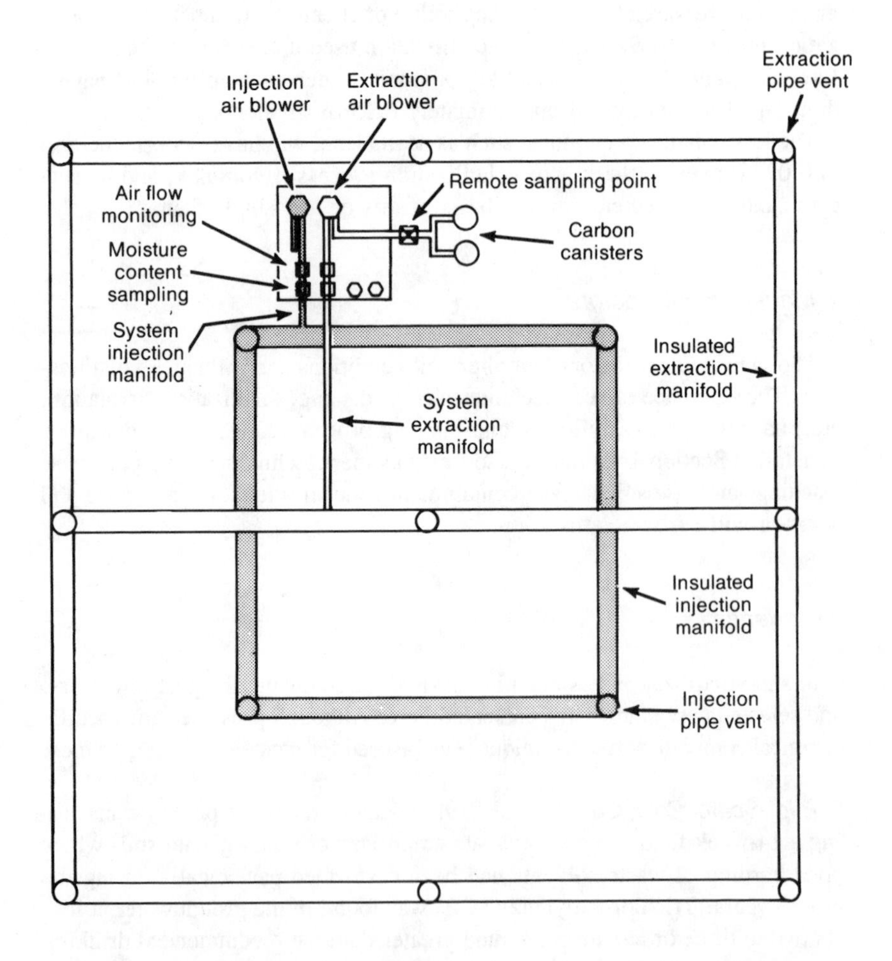

Figure 3.3 Vent pipe and manifold system layout.

Table 3.2 Results of the Pilot Study for In Situ Volatilization of TCE

Pilot System No. 1 (Low TCE Concentrations)	Pilot System No. 2 (High TCE Concentrations)
Initial TCE concentrations in the soil ranged from 5 to 50 mg/kg (ppm).	Initial TCE concentrations in the soil ranged from 50 to 5,000 mg/kg (ppm).
Daily exhaust air TCE concentrations exhibited decreasing trends over a three-month period from a high range of 5–12 mg/L (ppm) early in the project to the low range of 500–800 μg/L (ppb) at the end.	Daily exhaust air TCE concentrations remained essentially constant in the range of 250–350 mg/L (ppm) during the three-month period.
Approximately 1 kg (2.2 lbs) of TCE was removed from the soils containing low-levels of TCE.	Approximately 730 kg (1,606 lbs) of TCE was removed from the highly concentrated soils.
TCE removals to below 100 μg/L (ppb) in the exhaust air may have been achieved through continued system operation.	Similar TCE removal rates may have continued until TCE mass in the soil was significantly reduced. Only an estimated 10 to 20 percent of the TCE was removed from the soil during the short-term pilot test. Continuation of this operation on a full scale could be expected to remove a high percentage of subsurface TCE.

Previous soil sampling and analysis performed at the site identified the most likely location of at least two areas containing TCE from past solvent burning activities. Pipe vent locations were based upon data from these initial soil borings. Two separate pilot systems were installed with a depth of 6.1 meters (20 ft) for the pipe vents at the two areas of testing. Pilot System No. 1 was designed to test volatilization of low-level TCE concentrations identified in the soil of the first burn area (TCE soil concentrations of 5-50 mg/kg (ppm)). Pilot System No. 2 was designed to test this volatilization technique within the soils of the second burn area which contained high levels of TCE (50 to 5,000 mg/kg). The exhaust air was routinely sampled to establish TCE removal rates. Soil samples were collected and analyzed before and after the in situ volatilization to provide an order of magnitude estimate of TCE removals.

Results of this study are presented in Table 3.2. The pilot-scale volatilization program demonstrated the successful application of this technology as an in situ treatment strategy for remedial action.

Full-Scale. Hoag and Cliff (1985) have documented the use of in situ volatilization for the site remediation of a service station. Approximately 1,500 to

1,900 liters (400 to 500 gal) of gasoline were spilled and penetrated the unsaturated zone to the groundwater table. Approximately 300 liters (80 gal) of free-floating gasoline were recovered by depressing the water table with groundwater pumping and skimming or bailing the free-floating gasoline.

Figure 3.4 shows a plan view of the service station and venting pipe scheme. Three 0.15 meter (6 in.) diameter wells were placed in the unsaturated zone with piping connections to three vacuum pumps. Figure 3.5 shows the cross section view of one of these well-to-vacuum pump systems. The vacuum pumps were rated for 0.6 m^3/min (21 cfm).

During the 90-day operating period, 1,376 liters (364 gal) of gasoline were recovered from the vapor phase using this volatilization technique. Ninety percent of the recovered amount or approximately 1,244 liters (329 gal) was removed in the first 40 days. Ten percent was recovered over the remaining 50 days. These rates of initially high withdrawal and progressively decreasing recovery were plotted by Hoag and Cliff and are reproduced in Figures 3.6 and 3.7.

Although recovery rates are dependent on the site-specific soil, environmental, chemical, and management parameters addressed earlier, results such as those described above dramatically demonstrate the potential for remedial action utilizing this in situ volatilization.

Full-Scale Positive Pressure System. In addition to the negative pressure systems previously described, positive pressure systems can also be used to establish and maintain a barrier to migration of subsurface vapors. Positive pressure systems do not provide appreciable site remedial activity, but may be required in certain situations. Used alone, or in conjunction with in situ volatilization, this technique minimizes the impact to subsurface structures and dwellings.

O'Conner, Agar and King (1984) document the use of this technique at the site of an old apartment building built upon alluvial gravels in the vicinity of several service stations. Gasoline vapors were noted in the basement and on the main floor of the apartment building, and an inspection revealed cracks in the basement walls and floor slab. Only partial patching of these cracks was possible due to access problems.

In order to limit business interruption to the commercial tenants a positive pressure was established in the basement using fans. The gasoline vapor concentrations were so dramatically reduced that normal operations were restored within hours. Subsequent investigation determined the source of the vapors and an in situ volatilization system was installed. During the one-year operation the system recovered sufficient product from the vadose zone to allow termination of the project.

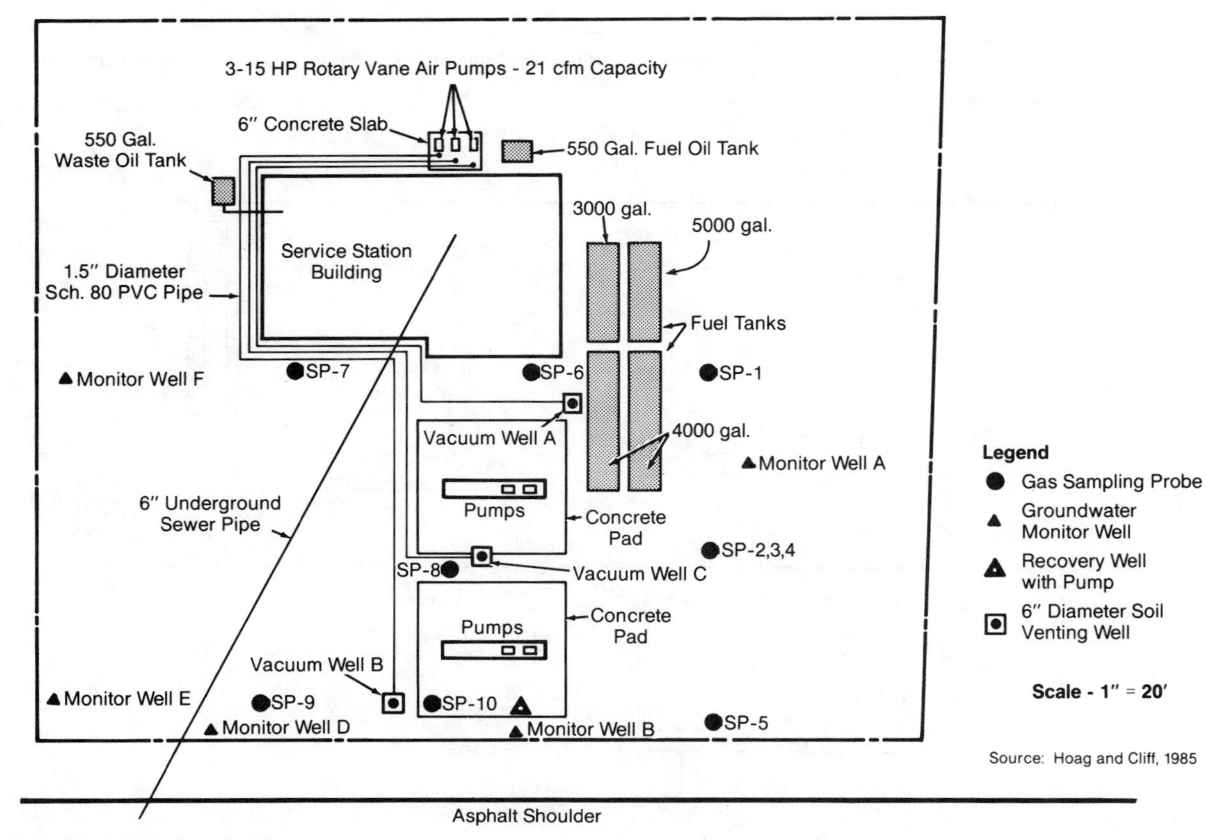

Figure 3.4 Service station site plan.

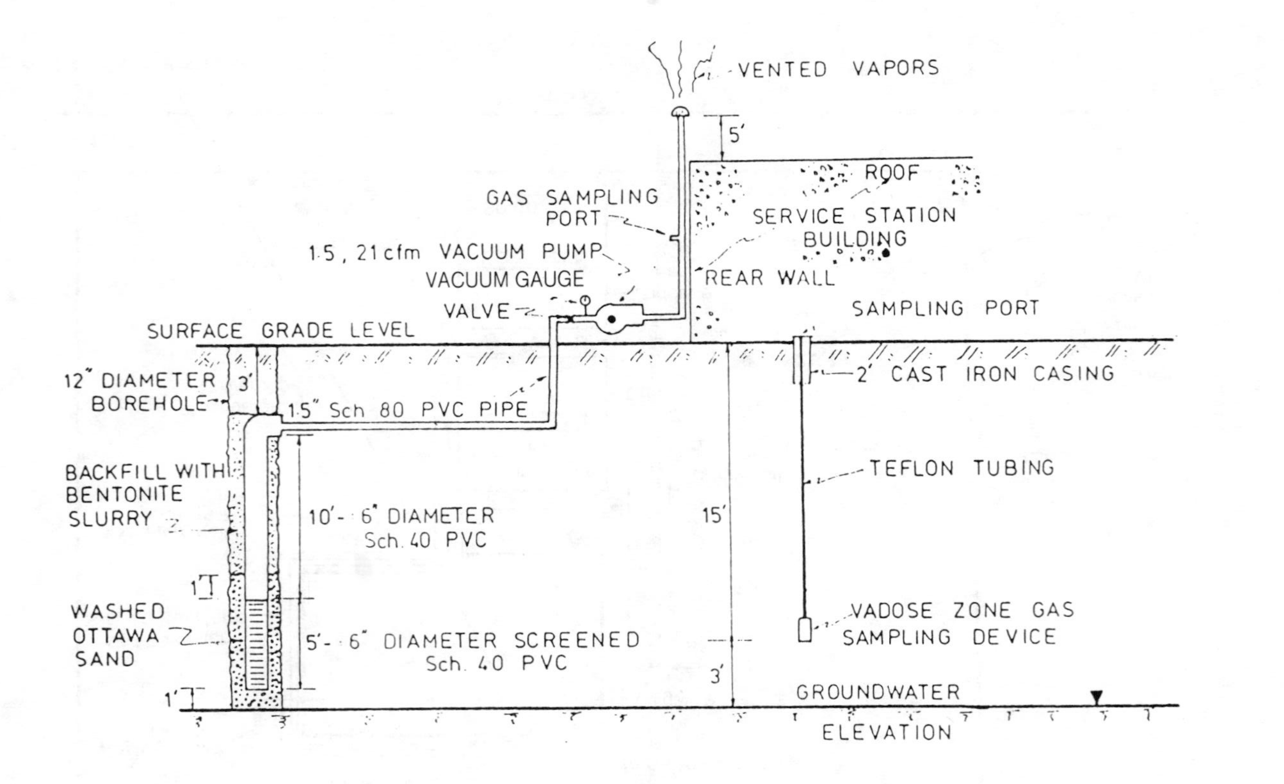

Source: Hoag and Cliff, 1985

Figure 3.5 Schematic plan for soil venting system with gas sampling probes.

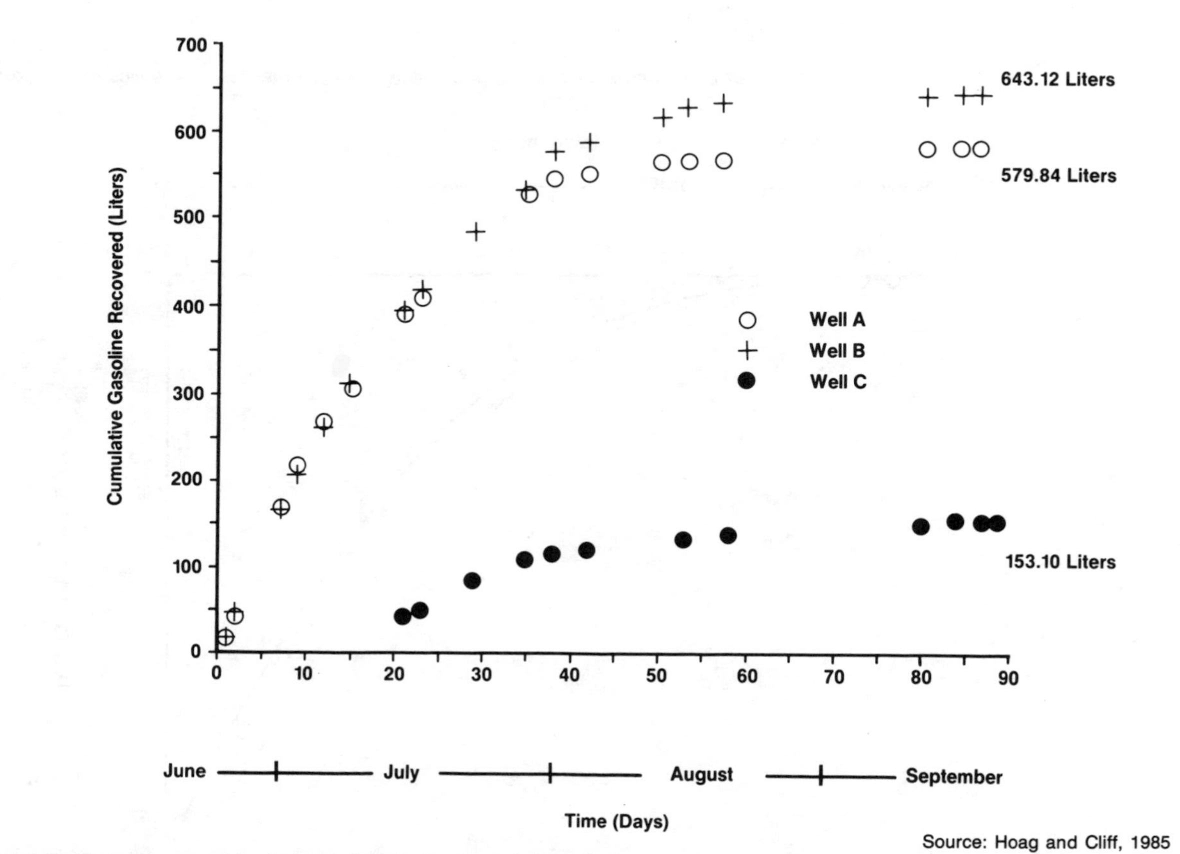

Source: Hoag and Cliff, 1985

Figure 3.6 Gasoline recovery during full-scale in situ volatilization.

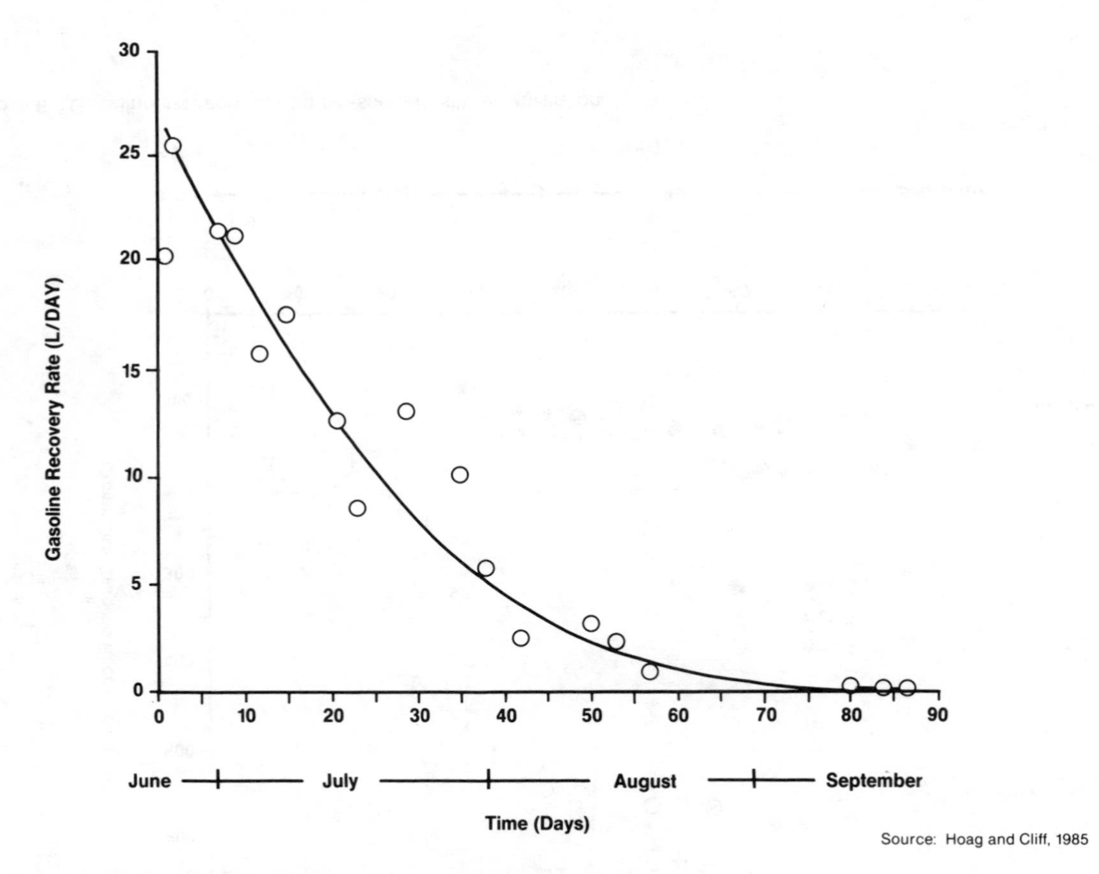

Figure 3.7 Gasoline recovery rate—Monitor Well A during full-scale in situ volatilization.

Implementation Feasibility of In Situ Volatilization

Design Considerations

The design of the in situ volatilization system must address all site-specific operational constraints. Design considerations that are common to in situ volatilization systems are as follows:

- Vent Spacing—The spacing of vents may range from 7.6 meters (25 ft) or less to 328 meters (100 ft) or more. Factors affecting vent spacing include flow rate, soil porosity/permeability and areal extent of hydrocarbon migration. A single vent or multiple vents can be used.
- Vent Depth—The depth of vent screening is a function of both the site geology (i.e., clay distribution) and the depth to which hydrocarbon compounds have migrated.
- Blower Requirements—The blower specification can be assessed by calculating expected total pressure drop through the system for the required air flow rate. Reference can then be made to the manufacturer's rating curves for blower selection.

In addition, certain site-specific conditions may also require heated or insulated piping and equipment and/or location of piping below or at the surface.

Equipment Requirements

The basic equipment requirements for implementation of in situ volatilization are listed below. All equipment should comply with National Fire Protection Association (NFPA) and National Electrical Code (NEC) guidelines. Considerable site-to-site variation is to be expected due to site-specific factors; however, key elements of most systems will be:

- Vent Ductwork—This can be any circular or rectangular ductwork, manifolded or not, with the structural integrity to withstand subsurface loading and with the chemical resistivity to assure sufficiently long lifetimes. Perforated PVC piping is often utilized for this purpose.
- Air Movers—Fans and/or blowers are used to move air through the ductwork and unsaturated zone. A wide variety of such units are available and one must be chosen on the basis of expected pressure drop, required flow rate, chemical resistivity (for induced draft), noise output and cost. Explosion-proof motors and non-sparking wheels for these units may be necessary. For large systems varying speed or multiple fixed speed blowers can be used.
- Air Flow Control—A system of ductwork manifolded to a single air mover

may require balancing to assure the proper air flow to each leg of the system. By adjusting flow control valves on vent piping, proper air flow throughout the system can be maintained or balanced. A series of air flow meters and flow restrictors can be used to accomplish this.

- Heaters—Air heaters are used to increase the temperature of the influent. Portable kerosene, propane and electric heaters can be used; however, electric heaters with portable generators are also commonly employed. Various heaters are used to prevent equipment and piping freeze-ups.
- Monitoring—To facilitate monitoring of air flow rates, air flow pressures and temperatures, an automated microprocessor data logging system can be used. In many cases, the installation of simple pressure gauges and flow meters is all that is necessary.
- Exhaust Emissions Control—Exhaust gases may require treatment prior to air discharge. A variety of treatment options are available.

Treatment Needs

The discharge of hydrocarbons to the atmosphere may be restricted by regulatory agencies. Discharge permits for direct venting or treatment prior to venting may be required. Applicable requirements are discussed under Permitting Requirements. However, should treatment be required, the following emission control system options are often considered:

- Vapor Phase Carbon Adsorption—This treatment method is widely used for the removal of organics from vent air streams and is well suited to emission control for dilute vent streams. Granular activated carbon can be regenerated or disposed of and replaced.
- Incineration—This treatment method is considerably more costly but results in the destruction of the organic constituents to carbon dioxide, water, and trace by-products. For dilute vent streams supplemental fuels will be necessary to maintain combustion temperatures. Cost-saving measures include converting an existing boiler to receive the vent stream or recovering the heat directly for steam generation. This option may be particularly effective if the site is located at a large generating facility where the vent stream may constitute only a small fraction of the existing combustion air usage. A variation of this treatment is the use of catalytic oxidation which significantly lowers the required combustion temperature by utilizing a compatible catalyst. In this process catalyst fouling can cause efficiency reductions, particularly if chlorinated organics are present.
- Flaring—This treatment is widely used in the petroleum industry for emergency venting. Flaring has recently been more widely investigated as a cost-effective continuous emission control measure. Lower heat value

gases can be flared successfully down to 8,928.6 Btu/scm (250 Btu/scf) without additional heat input.

Utilizing this treatment scheme for the very dilute vent streams produced by in situ volatilization would require supplemental fuel; hence, the treatment becomes much less cost effective.

Disposal Needs

In situ volatilization that does not require treatment of the gaseous emissions produces no waste or by-products requiring disposal. However, disposal options may be required should carbon adsorption be utilized as a treatment for the exhaust gases. Spent carbon will require disposal in an approved landfill or may instead be regenerated. Existing facilities are currently available for either disposal or regeneration. Consideration of proximity to the disposal/regeneration facility, cost of option, lifetime of carbon bed, and cost of fresh carbon will help determine the choice of option.

Monitoring Requirements

Site and process parameters require monitoring to verify process efficiency and ensure containment of chemical compounds for any remedial activity. The particular monitoring plan chosen will depend on site-specific factors and will most likely be negotiated with regulatory agencies on a case-by-case basis. The following parameters should be considered for monitoring:

- Groundwater—Monitoring will reveal if compounds are migrating and, if so, whether groundwater remediation should be initiated. At least one well should be placed upgradient of the petroleum source in a region with background levels of target hydrocarbons. Downgradient wells should be placed in order to detect downgradient migration; the number of downgradient wells will depend on the size of the site. Hydrogeologic conditions that influence flow velocity will be measured and used to evaluate sampling frequency.
- Soils—Periodic evaluation of hydrocarbon concentrations in the soils can provide an order-of-magnitude estimate of hydrocarbon mass remaining in the vadose zone. Samples should be taken in a number of locations ranging from the most affected soil locations to background regions.
- Process Air Flows and Temperatures—Periodic monitoring will ensure proper system balance and provide a data base which can be used to evaluate process changes.

- Exhaust Gas—Monitoring of exhaust gas before and after treatment may be required by operating permit to verify levels of air emissions. Monitoring will provide feedback on treatment efficiency and, in the case of carbon adsorption, reveal the need for carbon exchange.
- Equipment—Monitoring and preventive maintenance on fans, instruments, and other equipment will help prevent unexpected emergency system shutdowns.

Permitting Requirements

Permits from several government and regulatory agencies may be required. On a local level, establishing an ISV unit may require compliance with building and land use ordinances. Treatment of the system exhaust gas may be required by state air pollution agencies. Table 3.3 presents selected state regulations for volatile organic compounds.

Table 3.3 State Standards for Volatile Organic Compounds

State	Emission Standard for Total VOCs
Indiana	1.36 kg (3 lbs)/hr 6.8 kg (15 lbs)/day
Oregon	Attainment areas—36.4 metric tons (40 tons)/yr Non-attainment areas—10.2 metric tons (20 tons)/yr
South Dakota, Texas, Utah, Arizona, others	36.4 metric tons (40 tons)/yr

Notes:

1. In many states, including those above, VOC emission is further regulated on an industry basis. Reference must be made to the specific state air pollution control laws and regulations.

   Regulation of air emission standards for specific compounds (such as benzene, toluene, and xylene) is often administered on a regional level within a given state. For example, California administers air emission standards for specific volatile organics through its regional air quality management districts.
2. State regulations may not address environmental cleanup operations which are generally addressed through state regulatory agencies on a case-by-case basis.

Environmental Feasibility of In Situ Volatilization

Exposure Pathways

During the operation and monitoring of an in situ volatilization system, the primary human exposure pathways that must be considered are vapor inhalation and skin contact. Particle inhalation or ingestion could also be a primary pathway during system installation when soils are disturbed, but would be considered a secondary pathway during system operation and monitoring when this pathway is minimized.

Impacts to groundwater and surface water are considered the primary environmental pathways because these impacts usually determine the cleanup levels at sites where groundwater and surface water are affected by the petroleum products.

The emission control systems described under Treatment Needs will minimize the transfer of compounds from the treated soil to the environment. Maintenance of an emission control system that utilizes carbon adsorption may involve additional exposure pathways. These additional pathways could result from vapor inhalation, particulate inhalation/ingestion, and skin contact with air and/or activated carbon that are more concentrated than the soil gas which is treated.

Environmental Effectiveness

The effectiveness of in situ volatilization techniques is highly dependent upon site-specific conditions. As stated earlier, soil porosity, clay content, ambient temperatures and a variety of other factors all influence the effectiveness of ISV. Full-scale, bench-scale and pilot-scale case studies generally confirm the following effects:

- In situ volatilization has been successful for remediation in an unsaturated zone containing highly permeable sandy soils with little or no clay. In soils with low porosity or clayey soils, additional time is needed to establish the pressure gradient required to enhance volatilization.
- Recovery periods have been as short as a few weeks and typically on the order of 6 to 12 months. As soil contaminant concentration decreases, the mass rate of recovery per unit time drops and the effect of diminishing returns may determine project completion.
- Lighter, more volatile, components of petroleum products such as those found in gasoline show the greatest recovery rates. In some cases, free product on the groundwater table has been virtually eliminated via in situ volatilization.

- In situ volatilization can be used in conjunction with product recovery systems. These systems utilize low temperatures or carbon adsorption to lower exhaust air treatment costs by recovering a valuable asset.
- Ultimate cleanup levels are site-dependent and cannot be predicted. These levels may be set by regulatory agencies.

Economic Feasibility of In Situ Volatilization*

Capital Costs

Capital costs can be relatively low for the basic ISV system incorporating vertically installed vent piping, conventional fans or blowers, and basic monitoring and control devices. Costs will be a function of design flow rate, size of piping, degree of automated monitoring, and, if necessary, effluent treatment required. Costs for effluent treatment (carbon adsorption, incineration) could raise the cost of this technology an order of magnitude or more. Blower cost ranges from $300 to $3,000 or more depending upon design flow rate and pressure drop. Slotted PVC piping for Schedule 40 0.15 meter (6 in.) diameter pipe will cost less than $82 per linear meter ($25 per linear foot). Unslotted PVC piping of similar size will cost less than $65.6 per linear meter ($20 per linear foot). PVC fittings (elbows, tees, etc.) with diameters greater than 0.15 meter (6 in.) may exceed $100 each. The cost of a monitoring device or data logger will be $4,000 or more depending upon the degree of sophistication desired.

Installation Costs

A system utilizing PVC piping is relatively easy to install due to the ease of pipe joining and low weight. Installation costs for vertically installed pipe vent are similar to that of groundwater monitoring wells since the same construction materials and techniques can be used. Installed costs range from $48 to $65.6 for a 0.15 meter ($15 to $20 for a 6 in.) pipe per meter (foot) of depth. The installation costs for exhaust gas treatment depend on the option chosen. Incineration options will raise the overall installation costs significantly.

---

*All costs are presented in 1986 U.S. dollars unless otherwise specified.

Operation and Maintenance Costs

Operation and maintenance costs are generally modest for these simple systems. The operation and maintenance costs will vary, largely due to the amount of automation which has been incorporated into the system. Operating costs are derived from power requirements for fan operation and vary with the cost of electricity. Annual cost for full-time operation can be estimated using the following formula:

Annual fan operating cost = Fan Brake hp × 0.746 kW/hp × 8,760 hrs/yr × Electrical Cost, $/kW-hr

Operations and maintenance costs for exhaust gas treatment may be high. Replacement of carbon periodically will be far less expensive than operation of even the simplest incineration unit.

In addition, costs for periodic inspection and data collection should be considered. Maintenance costs can be conservatively estimated to be 4 percent of total installed costs.

Qualitative Ranking of Cost

In situ volatilization systems, in their most basic form, rank low in terms of cost. Exhaust gas treatment systems, however, can raise the cost of the entire system dramatically. Depending on flow rate, carbon usage, and other factors detailed earlier the cost for a carbon adsorption system may still be modest. Incineration options can prove very costly, particularly when the presence of chlorinated hydrocarbons require more expensive construction materials.

IN SITU BIODEGRADATION

General Description

In Situ Biodegradation is the process by which the growth and activity of naturally occurring microorganisms are stimulated in their natural environment. These microorganisms, through their metabolic processes, degrade the compounds of interest.

Stimulation of microbial growth and activity for hydrocarbon removal is accomplished primarily through the addition of oxygen and nutrients. Several factors influence the rate of this growth, including temperature and pH. These factors, along with the feasibility of remediation based upon biodegradation, are discussed in detail in this subsection.

Biodegradation has been found to be an efficient and cost-effective method for the reduction of hydrocarbons in soils and groundwater. The technology is based upon that which has been succcessfully used for land treatment of refinery waste. Consequently, biotreatment of petroleum hydrocarbons in soils is well documented, and significant background information is available.

Process Description

Figures 3.8 and 3.9 depict typical systems for remediation by biodegradation. The basic operations are (Heyse et al., 1986):

- A submersible pump transports groundwater from a recovery well to a mixing tank.
- Nutrients such as nitrogen, phosphorus and trace metals are added to the water in a mixing tank. These nutrients are then transported by the water to the soil to support microbial activity.
- Hydrogen peroxide is added to the conditioned groundwater from the mixing tank just prior to reintroduction to the soil. As hydrogen peroxide decomposes it provides the needed oxygen for microbial activity. Hydrogen peroxide is also added directly to wells and infiltration galleries. Air spargers are sometimes operated in wells in lieu of peroxide injection.
- Groundwater is pumped to an infiltration gallery and/or injection well which reintroduces the conditioned water to the aquifer or soils.
- Groundwater flows from the infiltration galleries or injection wells through the affected area and then back to the recovery wells. The flow of the water should contact all soils containing degradable petroleum hydrocarbons.
- The water is drawn to the recovery well and pumped to the mixing tank to complete the treatment loop.
- Groundwater in which hydrocarbon concentrations have been reduced to very low levels is often sent through a carbon adsorption process for removal of the residual hydrocarbons. Biodegradation is less efficient at low substrate concentrations and would require longer treatment times. See Section 4, Solidification/Stabilization for more details on carbon adsorption and the use of other compatible technologies for treatment of the affected groundwater.

Alternatively, the groundwater may be treated biologically on-site at the surface. These treatment technologies are discussed in Section 4.

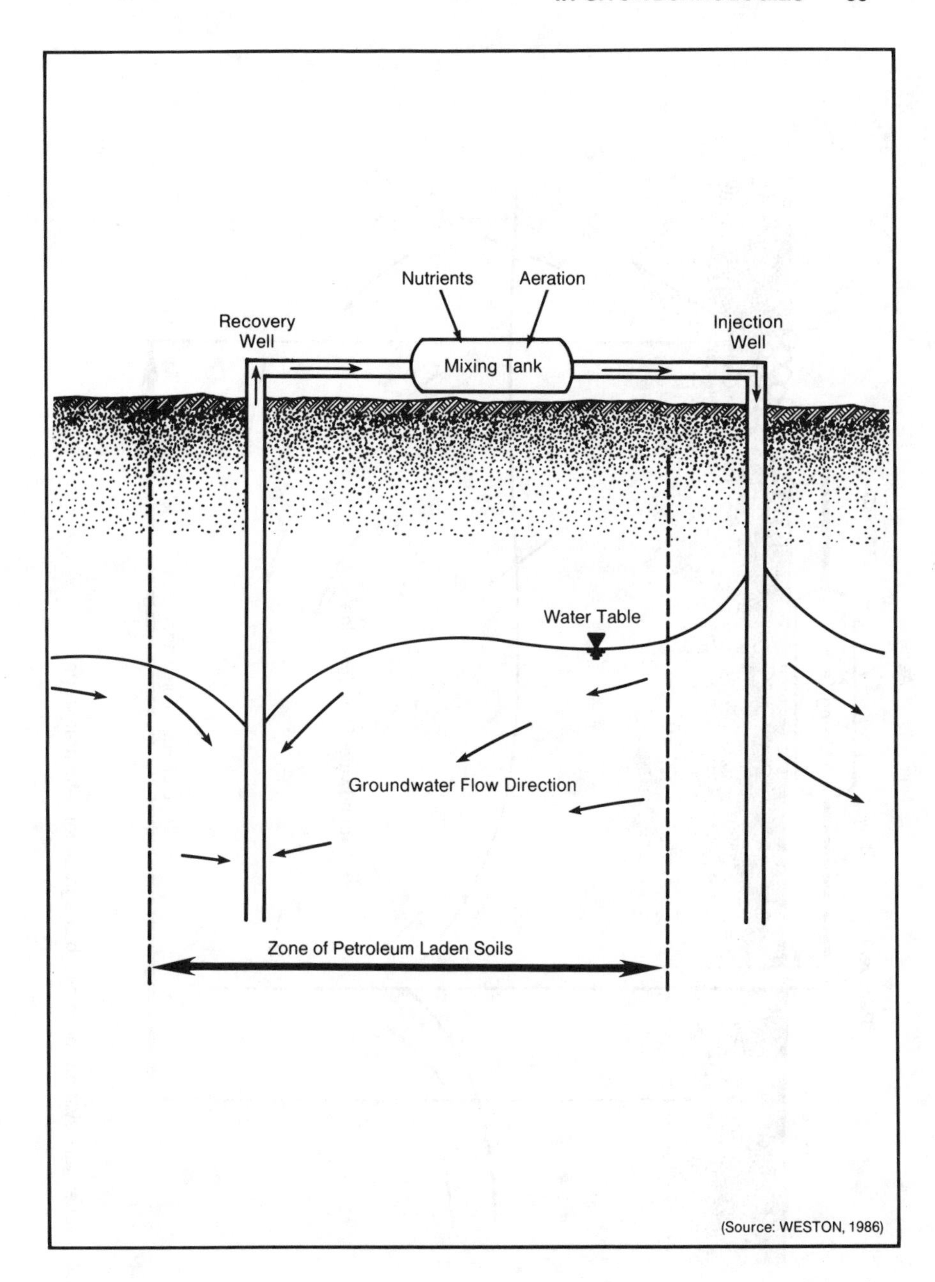

Figure 3.8 Schematic system for in situ biodegradation—injection well.

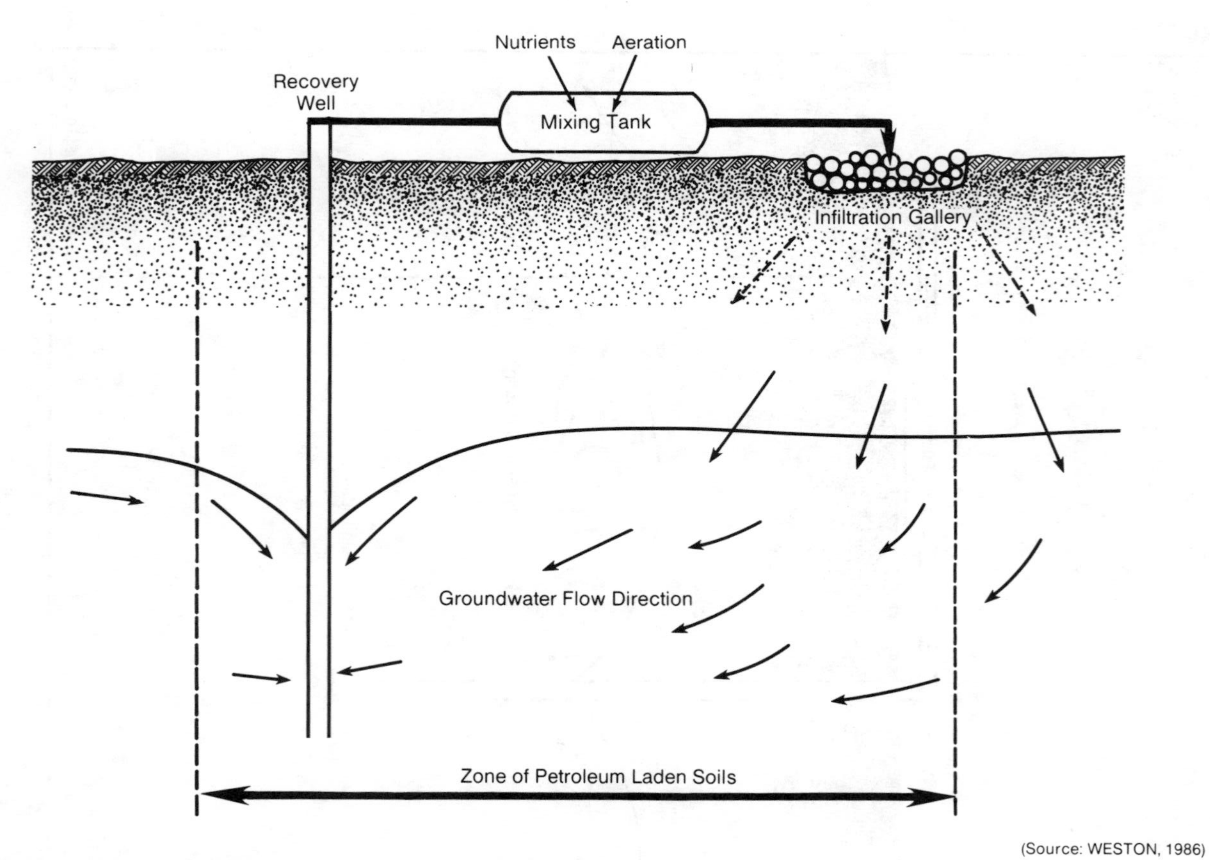

Figure 3.9 Schematic system for in situ biodegradation—infiltration gallery.

Technical Feasibility of In Situ Biodegradation

Technical Description

Biodegradation is a general term referring to the biological transformation of an organic chemical to another form (Grady, 1985). In situ biodegradation processes enhance natural biological activity in soil and groundwater in order to increase petroleum hydrocarbon decomposition. Most petroleum hydrocarbons can be degraded to carbon dioxide and water by microbial processes.

Hydrocarbon Biodegradation in Soils. Hydrocarbons are organic carbon compounds containing only carbon, oxygen, and hydrogen. A wide variety of bacteria, several molds and yeasts, and certain cyanobacteria and green algae have been shown to aerobically oxidize aliphatic hydrocarbons (Brock et al., 1984).

Many factors related to compound structure affect the biodegradation potential of specific hydrocarbon compounds (Bossert and Bartha, 1984; Bartha and Atlas, 1977; Atlas, 1981; National Academy of Sciences, 1984). In general, n-alkanes, n-alkylaromatic, and aromatic compounds of the C^{10}–C^{22} range are the least toxic to microorganisms and the most readily degradable compounds found in petroleum products. Compounds of this type, in the C^{5}–C^{9} range, tend to be easily removed by volatilization and can be toxic to microorganisms in high concentrations. Gaseous alkanes, in the C^{1}–C^{4} range, are biodegradable by a small group of specialized organisms. n-Alkanes, alkyaromatic, and aromatic compounds above C^{22} exhibit low toxicity, but their physical characteristics, low solubilities, and their tendency to remain in the solid state at normal temperatures combine to limit their biodegradability.

Branched alkanes and cycloalkanes of the C^{10}–C^{22} range are generally less biodegradable than their n-alkane and aromatic counterparts. Branching creates multiple carbon bonds that interfere with biodegradation. Highly condensed aromatic and cycloparaffinic compounds with four or more rings are largely resistant to degradation processes (Bossert and Bartha, 1984).

Factors Affecting Biodegradation in Soils. The biodegradation of a compound under field conditions is affected by many factors, including dissolved oxygen levels, soil moisture content, soil permeability, oxidation-reduction potential, temperature, pH, compound availability and concentration, availability of nutrients, and the natural microbial community (Table 3.4). Because these factors act together to determine the biodegradability of a compound in a particular setting, it is difficult to make generalizations about the importance of any one factor.

Table 3.4 Factors That Influence Biodegradation in Soils

Soil Factors	Environmental Factors	Chemical Factors	Management Factors
Moisture Content	Temperature	Type of Hydrocarbon Available	Availability of Nutrients
pH	Microbial Community	Concentration	
Oxidation/ Reduction Potential			
Porosity/ Permeability			

Notes:

1. It is extremely important to have a microbial community present that is capable of degrading the target compound. Most in situ biodegradation schemes make use of existing microbial populations; however, attempts have been made to supplement these populations with additional organisms or engineered organisms.
2. Chemical concentration has an important impact on biodegradation capability. Removal of any recoverable product that saturates the pore spaces or floats on the water table is recommended because hydrocarbons in the pure phase are degraded at a significantly lower rate.

Temperature—Degradation of petroleum products can occur at a wide range of temperatures. Huddleston and Cresswell (1975) reported petroleum biodegradation at temperatures as low as −1.1°C (30°F). The highest observed degradation rates appear to occur at temperatures ranging from 30 to 40°C (86 to 104°F) (Bossert and Bartha, 1984). In general, biodegradation of petroleum fractions increases as temperatures increase due to increased biological activity.

Chemical Factors—The availability and type of target compounds (or substrate) is a very important factor affecting the extent of biodegradation. Biodegradation is usually limited by the solubility of a compound in water, as most microorganisms inhabit soil moisture or need moisture in order to acquire nutrients and avoid desiccation. The substrate concentration is also an important factor as the growth rate of an organism is affected by the amount of substrate present. If the concentration is too low, the compound may not be metabolized by a microbial population which may favor another substrate that is available in higher concentrations. On the other hand, very high substrate concentrations may be toxic to the microbial community.

Soil Factors—Aerobic conditions are required for the degradation of hydrocarbons. Reduced soil oxygen levels lead to sharply reduced hydrocarbon utilization by microbes (Huddleston and Cresswell, 1975). Although soil moisture is a necessary requirement for microbial life, saturated soils that are low in dissolved oxygen can hinder biological activity. Moisture content between 50 and 80 percent of the water-holding capacity is considered optimal for aerobic microbial activities (Bossert and Bartha, 1984). Because oxygen transfer is a key factor in in situ biodegradation processes, the soils must be fairly permeable to allow this transfer to occur.

The pH of soils can vary widely, but most soils are somewhat acidic (Bossert and Bartha, 1984). The soil pH directly affects the microbial population supported by the soil. Biodegradation is usually greater in slightly basic (or alkaline) environments at a pH of approximately 7.8. Degradation of alkanes was found to be minimal in acidic soils with a pH of 3.7 (Dibble and Bartha, 1979).

Nutrients in the proper amounts are required for optimal biodegradation of petroleum hydrocarbons. Most notable in the case of soils containing hydrocarbons are nitrogen and phosphorus. Petroleum usually contains high amounts of carbon, but limited amounts of nitrogen and phosphorus. Several workers have reported that the addition of nitrogen and phosphorus salts immediately stimulated oil biodegradation in soils (Kincannon, 1972; Verstraete et al., 1976; Dibble and Bartha, 1979), while others found the same treatment to have little or no effect (Jobson et al., 1974; Raymond et al., 1976; Lehtomaki and Niemela, 1975). These conflicting results point to the fact that other site-specific factors, such as pH, temperature, oxygen availability, and/or water levels may limit biodegradation (Bossert and Bartha, 1984). For example, Verstraete et al. (1976) reported that the addition of nitrogen and phosphorus was beneficial only after the low pH of the natural soil was adjusted upwards.

Experience

In situ enhanced biodegradation of hydrocarbons present in soils and groundwater is gaining more attention as a potential remedial action. Additional site application and monitoring data will be needed to compile a performance data base for this technology, particularly with respect to ultimate cleanup levels which can be attained. The following case studies involve gasoline migration through the vadose zone to the groundwater and treatment methods associated with cleanup.

Case Study 1—Leaking Underground Storage Tank. Yaniga and Smith (1985) reported a leak of an undetermined amount of gasoline from an underground storage tank. This leak resulted in concentrations of 10 μg/L (ppb) to 15 mg/L (ppm) in the groundwater.

A system was designed to treat both soils and groundwater. Groundwater was pumped through a central well, passed through an air stripper to remove volatiles and add oxygen, fortified with nutrients, and sent to an infiltration gallery at a rate of 113,400 to 132,300 liters (30,000 to 35,000 gal) per day. A supplemental treatment system involving neutralization, carbon absorption, ion exchange, and biodegradation was implemented for the removal of inorganics and residual dissolved organic compounds.

Supplemental oxygen and nutrients were delivered to the groundwater via observation wells. Spargers were initially used for aeration, but were often blocked due to biologic growth. Hydrogen peroxide was then added to the infiltration gallery and the former air sparging wells to enhance oxygen transfer.

After 11 months, a 50 to 80 percent reduction in organics was noted. Samples from several wells within the hydrocarbon migration route contained less than detectable concentrations of organic compounds.

Case Study 2—Tank Leak. Leakage of an unknown quantity of gasoline from a storage tank led to vapor accumulation in the basements of two restaurants in La Grange, Oregon (Lee and Ward, 1984). The water table was located 3.96 meters (13 ft) below the surface and was overlain by various soil types (large cobbles, clay, coarse sand, silt, gravel and topsoil).

Laboratory studies concluded that the primary nutrients required to stimulate the natural microbial population to degrade gasoline were ammonium chloride and sodium phosphate. Secondary nutrients required were iron, manganese, magnesium, and calcium. Groundwater would be circulated through the site by pumping from recovery wells, thereby drawing down the static water level 1.83 to 3.05 meters (6 to 10 ft) and reinjecting at trenches located upgradient of the tank.

Three recovery wells were installed at places where maximum gasoline concentrations in groundwater occurred. The groundwater was pumped out, nutrients added, and sent to the trenches. Air was supplied at the bottom of the trenches and diffused by layers of porous stone.

Product recovery amounted to 12,345.5 liters (3,266 gal) of gasoline. It took only a few days for the nutrients to circulate to the recovery wells, but almost 5 months elapsed before elevated oxygen levels were detected there. This phenomenon is typical of in situ bioreclamation systems and reflects the fact that oxygen is utilized at a greater rate by the microorganisms. Bacterial populations increased to six million times the original level.

After nine months, gasoline levels in the most heavily laden areas ranged from 100 to 500 mg/L (ppm) and dissolved organic carbon was approximately 20 mg/L (ppm). At the end of the year-long program, dissolved organic carbon ranged from 2 mg/L (ppm) to 5 mg/L (ppm). No gasoline vapors were detected in the restaurant basements. A groundwater monitoring program was implemented to ensure that the process was successful and the gasoline had degraded.

Implementation Feasibility of In Situ Biodegradation

Design Considerations

Design considerations for enhanced biodegradation processes include examination of the site hydrogeology, compounds to be degraded, other site conditions, and process operations.

Site Conditions. Site hydrogeology is the most important consideration for designing a biodegradation system. The overall goals of in situ bioreclamation systems require the following hydrogeologic controls:

- Hydraulic control of injection and recovery of water at the site must be achieved in order to ensure that hydrocarbons are not spread and are effectively recovered.
- The bioreclamation system must achieve an even distribution of water flow through the area to be treated. Soils containing silts or clays are not as favorable for bioremediation due to their relative impermeabilities.

The following conditions of site hydrogeology should be considered when designing an in situ biodegradation process (Roux, undated):

- Unsaturated zone (vadose)—Knowledge of soil characteristics assists in tracking hydrocarbon migration and adsorption onto soil particles.
- Aquifers—The hydraulic relationships between multiple subsurface aquifers assists in evaluating the migration of petroleum hydrocarbons. Monitoring wells can then be placed in proper strategic locations.
- Fluctuations—Leaks of petroleum hydrocarbons can lead to the formation of a separate phase or organic layer which floats on the water table. Seasonal or daily water table fluctuations will repeatedly wash subsurface soils with the immiscible hydrocarbon phase.
- Groundwater flow—Assessment of the horizontal and vertical components, as well as the rate of groundwater flow, can provide information on the extent of migration of leached materials from the source.

In addition to these hydrogeologic parameters, the site conditions which affect the actual biodegradation of petroleum hydrocarbons must be considered. These are listed in Table 3.4 and discussed under Technical Description.

Process Operation. Knowledge of site hydrogeologic conditions is crucial in evaluating the scenario of groundwater circulation for in situ biodegradation. By understanding the considerations previously mentioned, recovery wells and injection ports can be placed for maximum effectiveness. Pumping rates can be properly adjusted to optimize the hydrogeologic conditions for maximum circulation. A groundwater hydraulic model is usually necessary to design the optimal placement and capacities of the recovery and injection wells and infiltration galleries. Considerations for process operations are:

- Recovery wells—The recovery or extraction wells should be placed so that as groundwater is pumped out of the aquifer, migrating petroleum hydrocarbons are drawn toward the well(s). Depending on the size of the site and the distribution of the compounds, more than one well may be needed to achieve containment of migrating compounds.
- Injection wells—After groundwater has been extracted by recovery wells and mixed with nutrients, it can then be reintroduced to an aquifer through injection wells. Generally, injection wells extend the entire length of the affected zone and induce water flow from injection to recovery wells. Depending on the size of the affected area, more than one well may be needed to accomplish total contact with the migrating hydrocarbons.
- Infiltration galleries—An infiltration gallery is a trench system or excavation that is backfilled with drainage rocks. Groundwater which is pumped from recovery wells and mixed with nutrients is sprayed over the rocks to disperse the water and provide increased aeration. The water then seeps into the ground making contact with the affected soils. The gallery must be located above the source or in an area of high concentrations of the petroleum compounds.
- Nutrient Addition—Nitrogen, phosphorus, and trace elements are needed for soil microorganisms to utilize carbon found in the petroleum products to produce new cellular material (Brown et al., 1986). These nutrients are added to pumped groundwater which is then reinjected into the aquifer.
- Aeration—Oxygen must be supplied to soil microorganisms so that biodegradation of hydrocarbons can occur. Aeration is generally provided by adding hydrogen peroxide to pumped groundwater just prior to reinjection or infiltration, or by air spargers located in the wells or holding tank.
- Pumping rate—The pumping rate of recovery wells and the injection rate of the injection wells depend upon the site-specific aquifer characteristics. A pump test and often a groundwater model are required to assess optimal pumping rates.

Equipment Requirements

The basic equipment requirements for implementation of in situ biodegradation vary from site to site due to site-specific factors. However, key components of most systems will be:

- Drilling rig—A drilling rig is required for installation of recovery, injection, and monitoring wells. Usually, a contractor is hired for well installation.
- Pumps—Submersible pumps are used in recovery wells to pump groundwater to the nutrient mixing tank. Standard centrifugal pumps may be used to transport the groundwater from the mixing tank to the injection well or infiltration gallery.
- Mixing tank—The mixing tank is temporary storage for the groundwater while it is mixed with nutrients.
- Sprinkler—A sprinkler system may be needed to spray treated groundwater over the rocks in the infiltration gallery.
- Air spargers—Aeration of groundwater by air spargers may be utilized.

Treatment Needs

Operational treatment needs for in situ biodegradation include the addition of nutrients and oxygen to the groundwater. The in situ biodegradation process is continuous, and, therefore, nutrients and oxygen must be provided regularly. Depending upon the levels of hydrocarbons which can be removed, additional treatment of groundwater for removal of organics or inorganics may be needed prior to potable use. Source control measures may also be necessary depending on treatment effectiveness of the system.

Disposal Needs

Disposal issues usually are not a concern for in situ biodegradation because the process is designed to be self-contained. Petroleum hydrocarbons are concentrated in the soil and treated groundwater which circulates through the site, and through biodegradation are transformed to innocuous forms. However, depending on any additional treatment needs (such as activated carbon treatment) at a particular site, disposal may be required.

Monitoring Requirements

Monitoring the in situ biodegradation process serves to 1) measure its effectiveness in containing petroleum hydrocarbon migration; 2) track the treatment effectiveness of the biodegradation process; and 3) optimize the hydrodynamics of the injection and recovery wells.

Containment of the hydrocarbons is evaluated by taking water samples from monitoring wells placed downgradient from the site and in unaffected zones of the aquifer. Aquifers above and/or below the affected zone should also be monitored.

To track the effectiveness of biodegradation, soil samples are periodically collected and analyzed for petroleum hydrocarbons of concern and nutrients. Analysis for the hydrocarbons of concern indicates whether biodegradation is proceeding at an acceptable rate. Nutrients are monitored to ensure that soil microorganisms are receiving proper nourishment.

In order to optimize the hydrodynamics of the well field, water levels are monitored periodically (at a minimum once a month). Pump tests should also be performed to test the efficiency of each well. A decrease in the well's efficiency, which could be caused by plugging of the well screen, could lead to decreased control of the hydrodynamics of the aquifer. The wells could then be rehabilitated. If rehabilitation of the wells does not increase their efficiency, new wells may be required to maintain hydraulic control.

Permitting Requirements

Permits from several government and regulatory agencies may be required. States may require permits for groundwater recovery and injection wells in addition to the above ground nutrient mixing tanks and peroxide system. On a local level, establishing an in situ biodegradation process may require compliance with land use ordinances.

Environmental Feasibility of In Situ Biodegradation

Exposure Pathways

During installation, operation, and monitoring of an in situ biodegradation system, possible primary exposure pathways include the following:

- Inhalation of vapors from soil or groundwater during system installation and monitoring and/or inhalation of emissions from the treatment unit during operations and monitoring.

- Contact of skin with soil or groundwater during installation and monitoring and with the treated streams during operation and monitoring.
- Impacts to groundwater and surface water which often determine the regulatory cleanup levels.
- Particle inhalation and/or ingestion during system installation.

Secondary pathways include particle inhalation and ingestion during operation and monitoring. Vegetation uptake during installation, operation, and monitoring must also be considered.

Additional exposure pathways must also be considered if additional treatment is required (e.g., exposure pathways due to air emissions produced or disposal of filters used by the bioreactor).

Environmental Effectiveness

The effectiveness of in situ biodegradation is dependent upon a number of site-specific factors such as nutrient availability, the size and type of microbial populations, aeration, pH of the soil, temperature, and ease of biodegradation of the substrate. The historical record for this technology is limited; however, several full scale case studies suggest the following:

- In situ biodegradation is most effective for situations involving large volumes of subsurface soils or groundwater. For situations involving smaller volumes of soil, land treatment or passive remediation are technologies involving biodegradation which may be more applicable (refer to In Situ Passive Remediation and Section 4, Land Treatment Technologies).
- Significant degradation of petroleum hydrocarbons has occurred in a time period as short as 2.5 months. Typical time periods, however, are in the range of 6 to 18 months (Brown et al., 1986).
- In situ biodegradation has most often been used for the remediation of groundwaters impacted by gasoline. Table 3.5 was developed by Wilson and McNabb (1983) as an estimate of the potential for biodegradation of several compounds in unconfined aquifers. Table 3.5 can be used as a crude guide to the potential effectiveness of in situ biodegradation processes for various classes of compounds. Under controlled conditions, biodegradation rates may be greater than those proposed by Wilson and McNabb, according to these authors, "These predictions are based on cautious extrapolation from the behavior of these compounds in other natural systems and on our admittedly limited experience with their behavior in the subsurface environment."
- Recent research suggests that there may be limited biodegradation of alkylbenzenes such as benzene and toluene under anaerobic conditions (Wilson et al., 1986). Research into the complexities of biodegradation is an

Table 3.5 Potential for Biodegradation of Selected Organic Pollutants in Unconfined Aquifers

Class of Compounds	Aerobic Water, Concentration of Pollutant, µg/L (ppb) 100	10	Anaerobic Water
Halogenated Aliphatic Hydrocarbons			
Trichloroethylene	none	none	possible*
Tetrachloroethylene	none	none	possible*
1,1,1-Trichloroethane	none	none	possible*
Carbon Tetrachloride	none	none	possible*
Chloroform	none	none	possible*
Methylene Chloride	possible	improbable	possible
1,2-Dichloroethane	possible	improbable	possible
Brominated methanes	improbable	improbable	probable
Chlorobenzenes			
Chlorobenzene	probable	possible	none
1,2-Dichlorobenzene	probable	possible	none
1,4-Dichlorobenzene	probable	possible	none
1,3-Dichlorobenzene	improbable	improbable	none
Alkylbenzenes			
Benzene	probable	possible	none
Toluene	probable	possible	none
Dimethylbenzenes	probable	possible	none
Styrene	probable	possible	none
Phenol and Alkyl Phenols	probable	probable	probable**
Chlorophenols	probable	possible	possible
Aliphatic Hydrocarbons	probable	possible	none
Polynuclear Aromatic Hydrocarbons			
Two and three rings	possible	possible	none
Four or more rings	improbable	improbable	none

Source: Wilson and McNabb, 1983

*Possible, probably incomplete.
**Probable at high concentration.

on-going process, and significant advances are continually being made.

- In soils, the remedial target level for in situ biodegradation could be in the low mg/L (ppm) level for total hydrocarbons. Groundwater levels may be less than 100 µg/L (ppb) (Brown et al., 1986).

Economic Feasibility of In Situ Biodegradation*

Capital Costs

Capital costs for in situ biological treatment are greatly affected by the method of aeration. For hydrogen peroxide injection, costs can range from $3,500 to $5,000 for groundwater pumping rates of 37.8 to 151.2 liters (10 to 40 gal) per minute. If air spargers are used, costs can range from $11,000 to $15,000 for pumping rates of 37.8 to 151.2 liters (10 to 40 gal) per minute (American Petroleum Institute, 1986).

Capital costs for a groundwater monitoring well with PVC casing (6-inch diameter) are approximately $32.8 to $65.6 per linear meter ($10 to $20 per linear foot).

In general, capital costs for in situ biological treatment depend on the site conditions and remediation level required. Total costs may range from $20,000 to $200,000.

Installation Costs

Installation of wells may have costs that range from $49.2 to $65.6 per meter ($15 to $20 per foot). PVC casings are the least expensive and stainless steel casings are the most expensive.

Operation and Maintenance Costs

Annual operation and maintenance costs for in situ biodegradation systems with air sparger aeration can range from $5,000 to $10,000 for groundwater pumping flow rates of 37.8 to 151.2 liters (10 to 40 gal) per minute. For the same flow rate, hydrogen peroxide injection systems may range from $3,500 to $15,000 per year. As pumping rates increase, operation and maintenance costs will also increase proportionally for hydrogen peroxide injection.

Annual sampling and analytical costs for groundwater monitoring based on 4 samples per year and 3 monitoring wells may cost approximately $5,000.

---

*All costs are presented in 1986 U.S. dollars unless otherwise specified.

Qualitative Ranking of Cost

The cost of in situ biodegradation is relatively low. Other than well installation, no major construction is needed and equipment requirements are minimal. Therefore, costs for the process are kept low, compared to other remedial technologies.

IN SITU LEACHING AND CHEMICAL REACTION

General Description

In situ leaching is the process by which in-place soils are flushed with water, usually mixed with a surfactant, in an effort to leach the compounds present into the groundwater. The groundwater is then collected, downstream of the leaching site, through a collection system for treatment and/or disposal.

Although the method of leaching is general, several methods are available for the collection system and several different surfactants can be used. These, as well as the feasibility of this process, are discussed in detail in this subsection. This process is sometimes described as ''extraction'' in the literature. In this document extraction refers to a process that is not carried out in situ and is described in Section 4, Chemical Extraction.

Case studies have shown that in situ leaching can be an effective removal alternative for hydrocarbons in soils; however, the technology for product surfactant separation is not yet available. This drawback must be considered when evaluating the costs of in situ leaching since both surfactant and waste disposal costs are considerable.

Process Description

The process of in situ leaching involves eluting or washing the petroleum constituents from the soil for recovery and subsequent treatment of the leachate. Depending on the physical nature of the petroleum products and soil, either water or water with a surfactant can be injected into the soil. For the most part petroleum products are not very soluble in water; therefore, a surfactant is usually added to reduce the interfacial tension between the petroleum constituents and the water. The hydrocarbons are thus dispersed into and thereby removed by the leachate.

The leachate is injected through the soil zone to a collection system. The collection system is usually a series of shallow well points or subsurface drains.

It is then pumped from the collectors to the surface for removal and recovery of the product. The remaining leachate, now relatively petroleum-free, can be recycled to the beginning of the process. Figure 3.10 is a general schematic of the soil flushing process (EPA, 1984a).

Technical Feasibility of In Situ Leaching

Technical Description

In the unsaturated zone, water and petroleum products can be retained by interfacial tension in the pore spaces and can be held as a thin sheath that coats each soil grain (Farmer, 1983). A high water solubility, a low soil-water partition coefficient, and a porous soil matrix aid in the effective removal of petroleum products from soils using the soil leaching technique. The soil-water partition coefficient (K_{oc}), a measure of the equilibrium between the soil organic content and water, is the leading factor in controlling the effectiveness of soil flushing. A low K_{oc} value indicates a favorable leaching tendency of the constituent from the soil.

Experience

A wide range of chemical additives has been considered as agents for flushing hydrophobic constituents from the soil. Considerable research and development efforts by the petroleum products industry to increase oil production have led to successful applications of enhanced oil recovery (EOR) techniques.

Case Studies—Bench and Pilot-Scale. The Texas Research Institute (1979, 1985) has performed laboratory and pilot-scale studies on surfactant enhanced gasoline recovery in sand. The 1979 laboratory study showed that a combination of commercial nonionic and anionic surfactants was effective in removing 40 percent of the residual gasoline from sand columns. The 1985 investigation assessed the effects of surfactants used to mobilize spilled gasoline found in the capillary fringe of a modelled sand aquifer. Three surfactant application techniques were studied: a single application by percolation through the sand, a multiple (daily) application by percolation, and a multiple (daily) application by direct injection. The multiple application approaches were found to be more effective than the single application in the removal of residual gasoline (76 percent and 83 percent for the multiple applications vs. 6 percent for

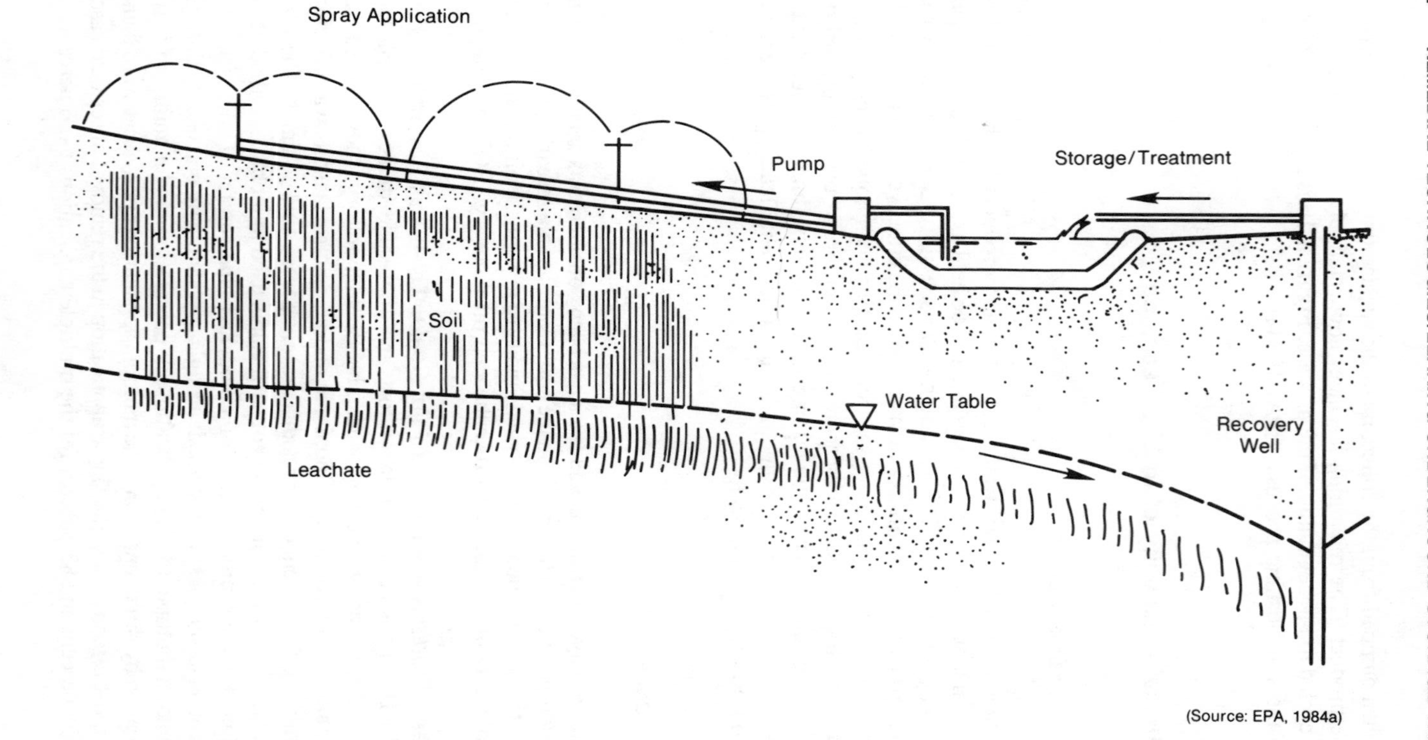

Figure 3.10 Schematic of a leachate recycling system.

the single application). In each case, the surfactant could not be separated from the gasoline constituents.

Ellis et al. (1985) studied treatment of soils containing PCBs and petroleum hydrocarbons with aqueous surfactants. Laboratory studies were carried out to develop an improved technique for the in situ treatment of soils bearing organic chemicals by soil flushing. An aqueous nonionic surfactant pair was used to treat soils in bench-scale shaker tests and in larger scale soil column tests. Ninety-two percent of the PCBs and 93 percent of the petroleum hydrocarbons were removed. These values are an order of magnitude greater than those observed when only the water flushing technique was being used. The authors also performed leachate treatability studies to investigate the potential for separation of the surfactant from the leachate. All attempts at removing the surfactant were unsuccessful.

Implementation Feasibility of In Situ Leaching

Design Considerations

In order to select the most appropriate delivery/recovery method and system, a thorough characterization of the site is needed. In general, information regarding the areal extent and vertical depth of the petroleum bearing soils, the surface and subsurface hydrologic characteristics of the site, and the geology and hydrogeology of the materials at and surrounding the site is necessary to decide upon the specific delivery/recovery system. Typically, the following information is needed:

- Extent and nature of the petroleum bearing soil.
- Site soil characteristics—porosity, permeability, uniformity, sorbtion potential, mineralogy, soil type.
- Surface drainage patterns and surface infiltration rates.
- Groundwater elevations, flow directions, and aquifer characteristics.
- Field permeability testing of the petroleum-bearing soils.

There are two general choices for delivery methods: forced and gravity. The forced delivery method can be used for all field conditions, while the choice of a gravity delivery method is more dependent on the above-listed characteristics of the site (Dul et al., 1984). Tables 3.6 and 3.7 present, in matrix form, information on several available gravity delivery methods and on the forced delivery method, with attention to various important factors at the site. Generally, when the petroleum-laden soils are located in the unsaturated zone where permeability is greater than $1 \times 10_{-3}$ cm/sec (.000394 in./sec), where there is a shallow permeable overburden and the depth to the bottom of the

Table 3.6 Delivery Methods for In Situ Leaching

Delivery Methods	Soils		Hydrocarbons Found at		Soil Deposit Topped by Layer of Impervious Material Having a Thickness			Topography		
	Unsaturated	Saturated	Surface	Subsurface	0M (0′)	1.5M (1.5′)	Above 1.5M (5′)	Flat	0–3%	3%
GRAVITY										
Flooding	X	NA	X	X	X	NA	NA	X	X	NA
Ponding	X	LE	X	X	X	X	NA	X	X	NA
Surface Spraying	X	NA	X	X	X	NA	NA	X	X	X
Ditches	X	NA	NA	X	X	X	NA	X	X	X
Infiltration Galleries	X	NA	NA	X	X	X	X	X	X	X
Infiltration Bed	X	NA	NA	X	X	X	X	X	X	NA
FORCED										
Injection Pipes	X	X	X	X	X	X	X	X	X	X

Source: Dul et al., 1984

X—Applicable
LE—Less Applicable
NA—Not Applicable

deposit is at most 4.6 meters (15 ft), gravity methods are quite adequate (Dul et al., 1984).

Recovery methods also may be classified as either gravity or forced. However, only two parameters need to be considered when making the preliminary selection of a recovery method. The first parameter of interest is the depth to the recovery zone. Gravity recovery methods are practical to a five-meter depth. For a deeper recovery zone, and for soils with a low composite permeability, forced methods are recommended (Dul et al., 1984). Table 3.8 presents information on the various types of recovery methods available and their suitability for specified field conditions.

Choice of surfactant type will also influence the effectiveness of the soil flushing process. Characteristics of surfactants and their environmental and chemical properties are found in Table 3.9 (EPA, 1985). This table can be employed as an aid in the preliminary selection of a surfactant. However, laboratory testing should be performed to quantify and verify surfactant/petroleum product interaction.

Table 3.7 Matrix for Delivery Methods

Delivery Methods	Infiltration Rate			Hydraulic Conductivity			Depth to Bottom of the Subject Soils		
	0.8 to 1.3 cm/hr	0.4 to 0.8 cm/hr	Below 0.4 cm/hr	10^{-4} to 10^{-3} cm/sec	10^{-3} to 10^{-4} cm/sec	10^{-4} to 10^{-7} cm/sec	4.6M	4.6 to 12.2M	12.2M
GRAVITY									
Flooding	X	X	NA	LE	NA	NA	X	LE	NA
Ponding	X	X	LE	X	LE	NA	X	LE	NA
Surface Spraying	X	X	NA	LE	NA	NA	X	LE	NA
Ditches	X	X	X	X	LE	NA	X	LE	NA
Infiltration Galleries	X	X	X	NA	NA	NA	X	X	NA
Infiltration Bed	X	X	X	NA	NA	NA	X	X	NA
FORCED									
Injection Pipes	X	X	X	X	X	X	X	X	X

Source: Dul et al., 1984

X—Applicable
LE—Less Applicable
NA—Not Applicable

Equipment Requirements

A mixing area or hopper for the preparation, processing, and storage of the surfactant/water solution must be provided in addition to a chemical feed system. If forced delivery methods are required, then a pump, or a series of pumps, with sufficient power will be needed to adequately force the leachate into and through the soil zone. A network of pipe diffusers or spray applicators will be needed for the delivery operation. Well points or subsurface drains with appropriate connecting piping must be installed to recover the leachate below the petroleum-laden soil horizon. The recovered leachate may be pumped up to the surface to be treated or collected for transport to another treatment or disposal facility. If the leachate is to be treated on-site, the treatment equipment for the chosen treatment operation must be supplied. The specific type of treatment equipment will depend on the type of process desired. The type of process is chosen based on the desired ultimate water quality results. Once

Table 3.8 Recovery Methods for In Situ Leaching

Recovery Methods	Depth to Recovery Zone 0 to 4.6M	4.6 to 12.2M	12.2M	Hydraulic Conductivity 10^{-1} cm/sec	10^{-1} to 10^{-3} cm/sec	10^{-3} to 10^{-4} cm/sec	10^{-5} to 10^{-7} cm/sec
GRAVITY							
Open Ditches and Trenches	X	NA	NA	X	X	NA	NA
Porous Drains	X	NA	NA	X	X	NA	NA
FORCED							
Well Point	X	X	NA	X	X	NA	NA
Deep Well	NA	X	X	X	X	NA	NA
Vacuum Well Point	X	X	NA	NA	NA	X	NA
Electroosmosis	X	X	X	NA	NA	NA	X

Source: Dul et al., 1984

X—Applicable
NA—Not Applicable

Table 3.9 Surfactant Characteristics

Surfactant Type		Selected Properties and Uses	Solubility	Reactivity
ANIONIC	Carboxylic Acid Salts	Good Detergency	Generally Water Soluble	Electrolyte Tolerant
	Sulfuric Acid Ester Salts	Good Wetting Agents		Electrolyte Sensitive
	Phosphoric and Polyphosphoric Acid Esters	Strong Surface Tension Reducers	Soluble in Polar Organics	Resistant to Biodegradation
	Perfluorinated			High Chemical Stability
	Sulfonic Acid Salts	Good Oil in Water Emulsifiers		Resistant to Acid and Alkaline Hydrolysis

Table 3.9, continued

Surfactant Type		Selected Properties and Uses	Solubility	Reactivity
CATIONIC	Long Chain Amines	Emulsifying Agents	Low or Varying Water Solubility	Acid Stable
	Diamines and Polyamines	Corrosion Inhibitor		
	Quaternary Ammonium Salts		Water Soluble	Surface Adsorption to Silicaeous Materials
	Polyoxyethylenated Long Chain Amines			
NON-IONIC	Polyoxyethylenated Alkylphenols and Alkylphenol Ethoxylates	Emulsifying Agents	Generally Water Soluble	Good Chemical Stability
	Polyoxyethylenated Straight Chain Alcohols and Alcohol Ethoxylates	Detergents	Water Insoluble Formulations	Resistant to Biodegradation
	Polyoxyethylene- and Polyoxypropylene-Glycols	Wetting Agents		Relatively Non-Toxic
	Polyoxyethylenated Mercaptans	Dispersants		
	Long-Chain Carboxylic Acid Esters	Foam Control		Subject to Acid and Alkaline Hydrolysis
	Alkylolamine "Condensates" Alkanolamides			
	Tertiary Acetylenic Glycols			
AMPHOTERICS	pH Sensitive	Solubilizing Agents	Varied (pH dependent	Non-Toxic
	pH Insensitivė	Wetting Agents		Electrolyte Tolerant Adsorption to Negatively Charged Surfaces

Source: U.S. EPA, 1985

treated, the surfactant/water mixture can be transported via another pipe to the head of the delivery area so that it may be reused in the soil flushing operation.

Treatment Needs

In order to have successful application of soil flushing in the field, large volumes of wash water and surfactants are necessary. To make the soil leaching process cost effective, recycling of the surfactants is an important consideration because of their relatively high cost. Several leachate treatment techniques have been evaluated for their ability to remove and concentrate the foreign soil substances while leaving the surfactant in the process water so that it may be reused. Ellis et al. (1985) report that all treatment methods evaluated (foam fractionation, sorbent adsorption, ultrafiltration, surfactant hydrolysis/phase separation, flocculation/coagulation/sedimentation, centrifugation, and solvent extraction) were found to be ineffective in separating the surfactant from the leachate constituents. Apparently, the chemical and physical properties of the surfactant/water mixture, desired for effective removal of soil constituents, are undesirable for the successful separation of surfactant. Until the technology is developed to the point where the surfactant can be actually separated from the product, reuse of the surfactant will be limited to that which has available sorption capacity.

In addition to devising a surfactant reuse phase of the treatment scheme, large volumes of leachate generated in the extraction process must be treated. Two additional objectives must be met: the eluates in the leachate must be concentrated to facilitate disposal; and the leachate water must be treated to an acceptable level for discharge to a POTW (publicly owned treatment works) (Ellis et al., 1985). Of the seven treatment methods mentioned above, only foam fractionation, sorbent adsorption, hydrolysis, and ultrafiltration successfully met these objectives when evaluated on a test-scale (Ellis et al., 1985).

Ultimate cleanup levels are both site- and constituent-dependent. These levels may be set by regulatory agencies. A fraction of the hydrocarbon/leachate will be left in the soil which may then require further treatment.

Disposal Needs

Because attempts to separate the surfactant from the leachate products have not been successful, the disposal of the leachate is an issue that must be addressed. When separation of surfactant and petroleum product is accomplished, the petroleum products recovered in the soil flushing operation must be processed or disposed of in order to prevent future release to the soil or nearby surface or subsurface water bodies. The process water separated from the surfactant and the petroleum constituents may also need treatment or pretreatment before ultimate disposal (either into a receiving stream, or as groundwater recharge, or to a POTW).

Monitoring Requirements

Because the process of soil flushing has not been implemented at large-scale sites, there have been no monitoring measures specifically required. However, to maintain effective operation, it is recommended that the operation be monitored at various points in the process system. For example, leachate flow should be measured at the delivery point and then compared to the flow from the collection pump in order to quantify discharge, if any, to the underlying saturated zone or groundwater. Monitoring wells, strategically placed in both the upgradient and downgradient flow directions, would be able to detect any surfactant that may have entered the saturated zone. One of the primary process control requirements is to assure that the surfactant solution can be effectively contained so that it does not result in spreading of the hydrocarbon. Also, it is important to monitor the amount of petroleum product recovered from the process to quantify the effectiveness of this technology.

Permitting Requirements

If the process water, upon completion of treatment, is to be discharged to a receiving stream, a National Pollution Discharge Elimination System (NPDES) permit is required under the auspices of U.S. PL 92-500, the Federal Water Pollution Control Act. Discharge to an existing on-site wastewater treatment plant may be possible and may require a permit modification. There may also be requirements for a groundwater discharge permit by Federal, state, and/or local agencies.

Regulatory agencies may be reluctant to issue approval permits for any activity that involves the injection of an additional chemical (i.e., a surfactant) to the subsurface. If the leaching and/or chemical reaction technology appears to be a viable alternative at a particular site, then extensive bench- or pilot-scale studies may be required before implementation approvals are granted.

Environmental Feasibility of In Situ Leaching

Exposure Pathways

The installation, operation, monitoring, and treatment of an in situ leaching system yields the following potential primary exposure pathways:

- Inhalation of vapors from surface water, groundwater, or the surfactant during system installation and monitoring.
- Inhalation of emissions from the above sources as well as the leachate itself during operations, monitoring, and treatment.
- Particle inhalation and ingestion during system installation.
- Contact of skin with surface water, groundwater, or the surfactant during installation, monitoring, operations, and treatment.
- Impacts to groundwater and surface water as primary pathways during installation, operations, monitoring, and treatment. The possible toxic nature of the surfactant being used must be addressed. The same environmental exposure pathways, groundwater and surface water, exist for the movement of surfactant as well as petroleum products.

Secondary pathways include dust inhalation, and particle ingestion during operation and monitoring when these impacts are minimized, and plant uptake during installation, operation, and monitoring.

Environmental Effectiveness

Due to the hydrophobic nature of most hydrocarbons, flushing with a surfactant is likely to be more effective than flushing with water.

Application of the surfactant flushing approach has not been undertaken at a major field site. Therefore, performance data on environmental effectiveness is unavailable. However, bench- and pilot-scale studies have produced results that range from 76 percent to 93 percent removal for specified constituents (Texas Research Institute, 1985; Ellis et al., 1985) (see the case studies discussed under Experience for details). Techniques still need to be developed to separate the surfactant from the soil constituents and the process water before cost-effective field application of this technology can be implemented (Ellis et al., 1985).

Economic Feasibility of In Situ Leaching*

Capital Costs

Capital cost data for in situ leaching and chemical reaction are scarce due to limited experience with these methods of remediation. However, costs for individual equipment needs can be addressed.

Submersible pumps (189 to 7,560 Lph (50 to 2,000 gph)) which are used in the wells normally cost between $400 and $1,000. Standard centrifugal pumps for the chemical feed system can range in price from $100 to $600. A sprinkler for the feed system may cost approximately $1,000. Capital costs for a 0.15 meter (6 in.) PVC well can cost $65.6–$82 per meter ($20 to $25 per ft).

In addition to capital costs for the in situ operation, costs can be incurred for treating the leachate to separate the surfactant and petroleum hydrocarbons.

Installation Costs

Installation of wells may have costs in the range of $49.2 to $65.6 meter ($15 to $20 per ft). PVC casing is the least expensive; stainless steel and Teflon casings are the most expensive.

Operation and Maintenance Costs

Annual sampling and analytical costs for groundwater monitoring based on 4 samples per year and 3 monitoring wells may cost approximately $5,000. Additional costs can be incurred during routine maintenance of equipment and the purchase of more surfactants. Costs for surfactants may range from $1.43 to $1.94 per kilogram ($.65 to $.88 per lb).

Qualitative Ranking of Cost

The cost of in situ leaching and chemical reaction is relatively inexpensive because the associated non-in situ tasks of excavation, transportation, and treatment are avoided.

---

*All costs are presented in 1986 U.S. dollars unless otherwise specified.

IN SITU VITRIFICATION

General Description

In situ vitrification is the process by which the in place soils are vitrified through the utilization of electricity. In the course of the vitrification process, the major portion of the compounds in the soils are volatilized with the remainder being worked in place in the hardened soil. The vitrification process, and its feasibility, is discussed in detail in this subsection.

Field pilot studies utilizing in situ vitrification have demonstrated that this process is very effective in removing a large amount of hydrocarbons present in the soil and also in containing the remaining hydrocarbons. This technology is relatively new and utilizes a large amount of electricity. These facts should be considered during the remedial technology selection process.

In situ vitrification is the conversion of soil into a durable glass and crystalline form by melting the soil by electrical heat. In situ vitrification is a new technology (first tested in 1980) and was originally developed for the purpose of stabilizing high level radioactive wastes in place. The electric melter technology required to implement this process was developed at Pacific Northwest Laboratories (PNL). A patent on the process is held by the U.S. Department of Energy (DOE), sponsors of the initial research (U.S. Patent No. 4,376,598; 1983). While this technology has been developed for soils containing radioactive materials, it may also be applicable to soils containing organic constituents (PNL, 1986; FitzPatrick et al., 1984).

Process Description

In order to initiate the in situ vitrification process, four electrodes must be inserted into the soil in a square pattern. Typical spacing for a large scale setting is from 3.5 to 5.5 meters apart. A small quantity of a mixture of graphite and glass frit is placed in an "X" pattern between the electrodes at the soil surface to provide a conductive path for the initial electrical current. As the electrical current is passed between the electrodes, the internal resistance of the conducting medium causes temperatures to rise, resulting in the melting of the adjacent soils. The soil becomes conductive as it melts, allowing the molten zone to continue to grow downward and outward. Less dense materials, such as rocks, migrate upwards to create a layer near the surface. Organic materials become pyrolized and diffuse to the surface, thereby creating a more porous zone near the surface. Ash or inorganic materials are encapsulated within the glass form. After the current is removed, one to two weeks are required for the vitrified mass to cool. During this time some subsidence occurs. Finally, the area can be backfilled and revegetated. Figure 3.11 details a view of the sequence of an in situ vitrification process (PNL, 1986).

As the molten mass grows, internal resistance decreases and the current must then be increased in order to melt more soil. To increase the current, power is supplied in stages using a power transformer with multiple power taps (PNL, 1986). The power system must be capable of performing at a number of conditions ranging from high voltage/low current to low voltage/high current. A large-scale system is designed to utilize a maximum power capacity of 3,750 kW with an average power of 3,200 kW. Buelt et al. (1985) provide a detailed description of the specialized power supply needed to perform the in situ vitrification operation.

An off-gas hood is included as part of the process in order to collect the off-gases, provide a chamber for combustion of pyrolized organics, and serve as a support for the four electrodes embedded in the soil (Buelt et al., 1985). Much of the heat that is generated by the vitrification process is released in the off-gas stream so that the final gas temperature will be nearly ambient. The off-gas stream can then be removed by a treatment system.

Technical Feasibility of In Situ Vitrification

Technical Description

In situ vitrification pyrolizes organics and immobilizes inorganic material by imposing an electrical current between electrodes placed in the soil. Heat generated by the applied electrical current can cause the soil matrix to attain temperatures greater than 1,700°C (3,092°F). A typical soil is primarily composed of silica and aluminum oxides which have melting temperatures between 1,100 to 1,600°C (2,012–2,912°F).

As soil is vitrified, organic constituents within the soil are pyrolized and the resulting gases combust at the surface when they contact air. High temperatures and long residence times result in essentially complete combustion and/or destruction of organic components (Buelt et al., 1985).

The resulting vitrified mass is similar in composition and weathering characteristics to obsidian, a natural glass-like material (Larsen and Lanford, 1978). Assuming a linear hydration rate, one can arrive at a conservative weathering estimate of a 1 mm hydrated depth for a 10,000-year period (FitzPatrick et al., 1984). Since hydration is the initial mechanism of weathering, the vitrified mass is expected to maintain its integrity for at least 10,000 years. Tests on vitrified soils have determined that the durability of the material is similar

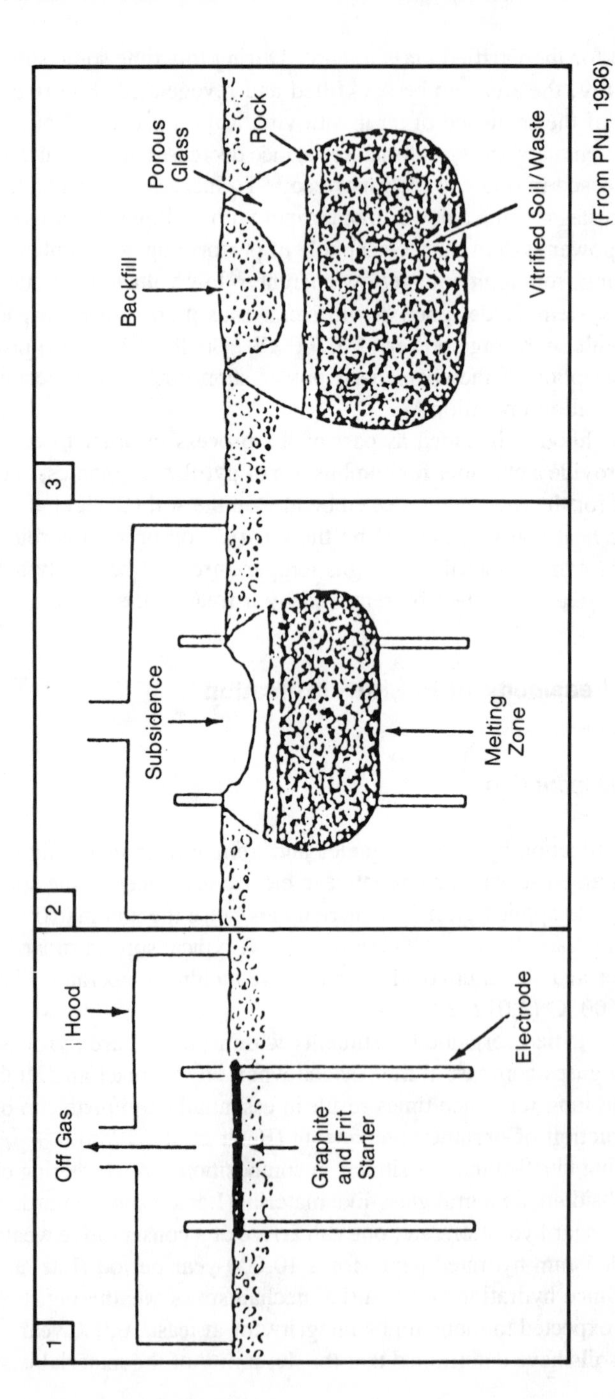

Figure 3.11 In situ vitrification process sequence.

to that of granite, and the leaching resistance is superior to that of marble and common bottle glass (Oma et al., 1983).

Experience

The in situ vitrification technique has been under development since 1980. While a number of research scale tests have been performed on radioactive soils under the direction of the DOE, this technique is also applicable to soils containing petroleum products. Table 3.10 summarizes the level of experience, as of 1985, at three levels of process development. Engineering, pilot, and large-scale testing of this technique have been performed at this time.

Case Study—Engineering Scale. Application of this technique to soils containing organic constituents is in the early stages of development. The Electric Power Research Institute (EPRI) has recently sponsored a successful engineering scale project at Pacific Northwest Laboratories (PNL) on the application of in situ vitrification of soils containing PCBs (PNL, 1986). The test was performed in a sealed container equipped with a vacuum controlled off-gas sampling system. A deep zone of loamy-clay soil, 0.2 meters (8 in.) in diameter by 3 meters deep (1 foot) contained 500 mg/L (ppm) PCB. This zone was centrally located between the electrodes under 10 inches of overburden. Electrodes were placed 0.23 meters (9 in.) apart and extended to a depth of 0.6 meters (24 in.). The resultant vitrified block weighed 480 lbs and had a volume of 0.14 cubic meters (5 ft^3). Figure 3.12 shows the engineering scale set up and the off-gas sampling system.

Results of the engineering scale test led to the conclusion that the in situ vitrification process has the capability of meeting current EPA standards regarding cleanup of PCB bearing soils. Conclusions regarding the test results are as follows (PNL, 1986):

- The small release of PCBs to the off-gas system (0.05 percent) can be effectively retained and treated by the appropriate design of a conventional activated carbon filter treatment system.
- Limited amounts of PCBs (0 to 0.7 mg/L (ppm)) were found in the surrounding soil, indicating that there is little compound migration during the in situ vitrification process.
- The in situ vitrification process resulted in a destruction and removal efficiency of greater than 99.99 percent, exclusive of off-gas treatment.
- Large-scale process tests are expected to range from $150.4 to $222.2 per cubic meter ($115 to $170 per yd^3) and are dependent on electrical power rates, soil moisture content, and labor rates.

Table 3.10 In Situ Vitrification Studies

Develop-mental Scale	Power Capability (kW)	Electrode Spacing (m)	Vitrified Mass Per Setting (metric ton)	Number of Tests Performed
Engineering	30	0.23–0.36	0.055–1.1	22
Pilot	500	1.2	11–55	10
Large	3,750	3.5–5.5	440–880	4

Source: PNL, 1986

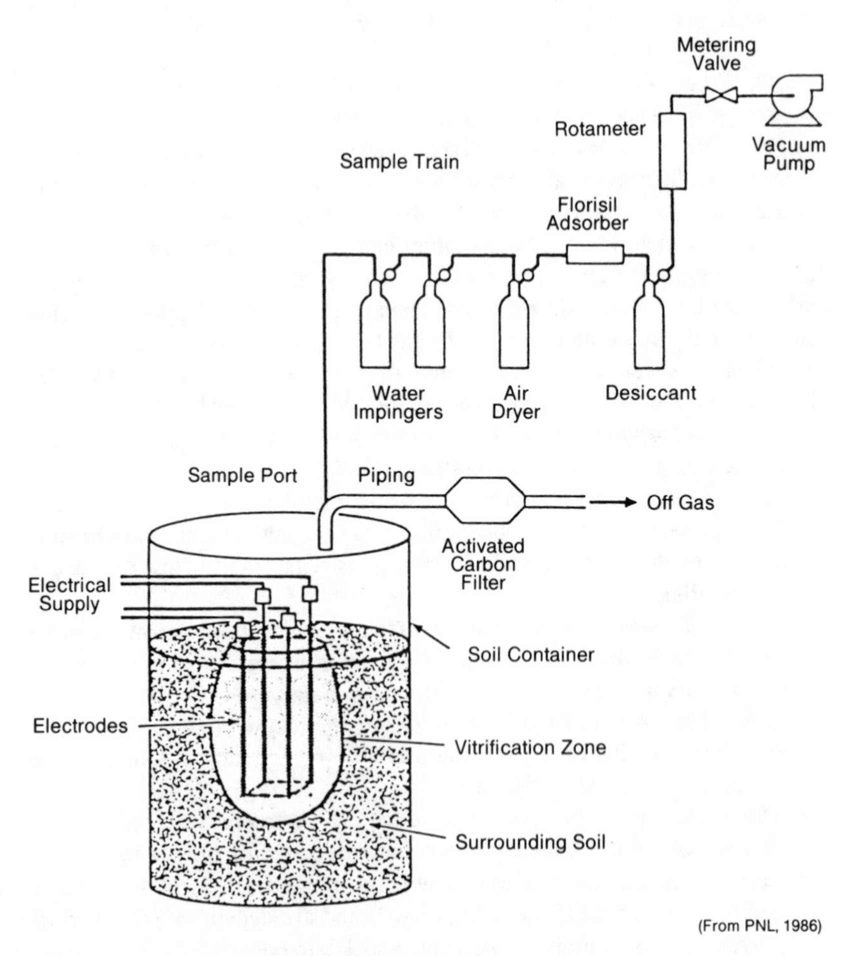

Figure 3.12 Engineering scale in situ vitrification system.

Implementation Feasibility of In Situ Vitrification

Design Considerations

Soils from various geographic locations in the U.S. have been vitrified in laboratory settings. Soil properties such as electrical and thermal conductivity, fusion temperature, and chemical composition do not significantly affect the vitrification process (FitzPatrick et al., 1984). High soil moisture content has a significant effect on the process resulting in higher power requirements and longer run times.

Tentative design parameters have been developed in accordance with the results of large-scale tests specifically targeted at the vitrification of landfills that contain mixed waste materials (FitzPatrick et al., 1984). These parameters do not necessarily reflect the limits of the system. Longitudinal metal shapes, such as pipes or bars, can short out opposing pairs of electrodes (Buelt et al., 1985). Test results for a typical electrode spacing width of 3.5 to 5.5 meters (11.5 to 18 feet) show that metals can occupy as much as 70 percent of the linear distance between electrodes with no effect on process performance. The size of void volumes (up to 4.3 cubic meters (153.57 ft^3)) and combustible packages (up to 0.9 cubic meters (32.14 ft^3)) are limited by the capabilities of the off-gas treatment system. Figure 3.13 summarizes the capabilities of a typical large-scale in situ vitrification system (FitzPatrick et al., 1984).

Releases to the off-gas system are inversely related to burial depth. A large proportion of combustible liquid can result in high gas releases. Other materials can be brought to the surface via entrainment.

Equipment Requirements

Equipment necessary to perform large-scale in situ soil vitrification is still under development. A large-scale system that would be available for rental on a commercial basis may be available by late 1987 (C. L. Timmerman, Personal Communication). Necessary system elements include the electrical supply system, the electrodes, the off-gas hood, and the off-gas treatment system. PNL, in conjunction with DOE, has constructed a mobile system used for large-scale testing purposes. All of the process equipment, with the exception of the off-gas hood, can be transported on three semitrailers.

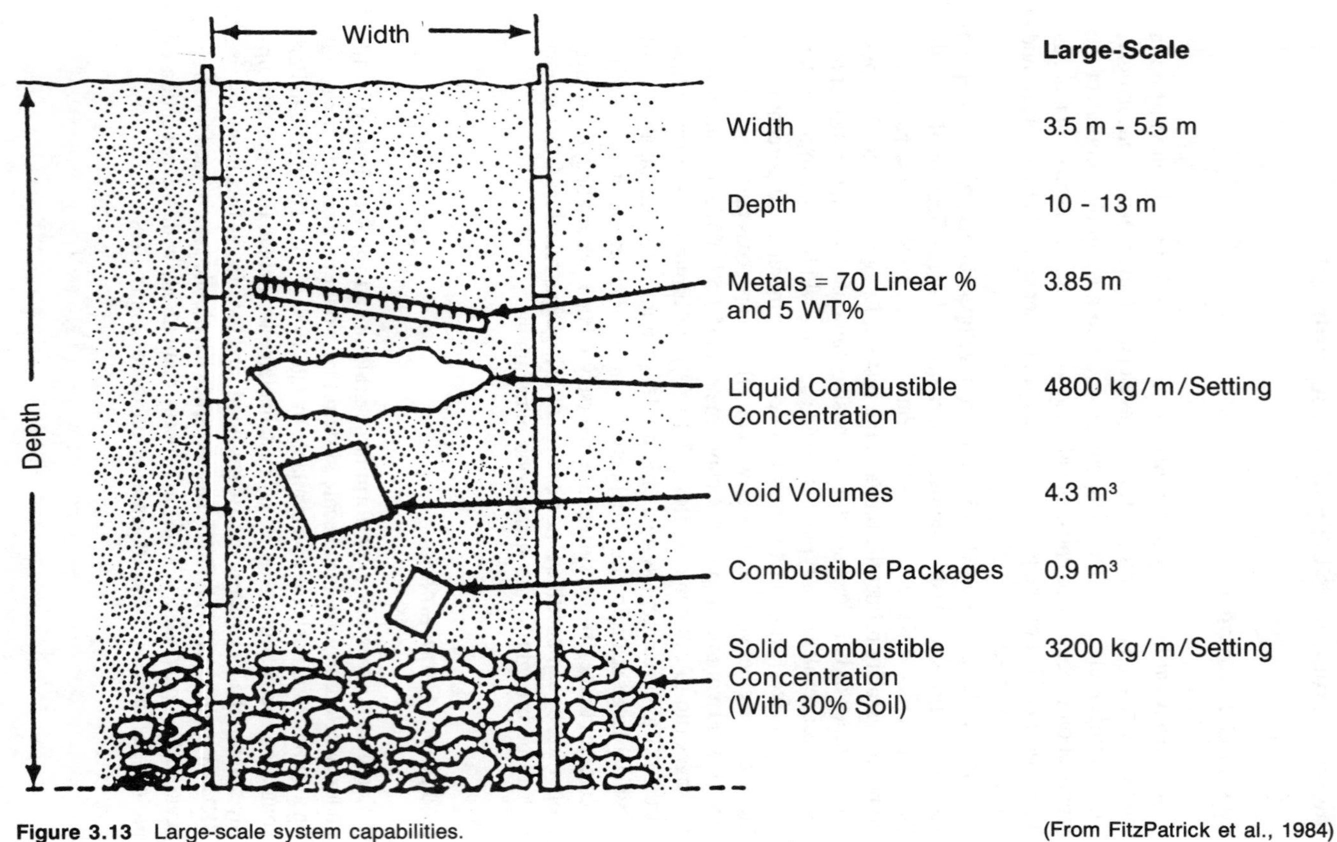

Figure 3.13 Large-scale system capabilities.

Treatment Needs

For treatment of hydrocarbons in soils, off-gas treatment is a major component in the overall system. A schematic of a large-scale off-gas treatment system is shown in Figure 3.14. There are three stages of treatment in such a system (Buelt et al., 1985). First, gases are cooled in a quencher and scrubbed in a tandem nozzle scrubber to remove particles down to the submicron range. Second, most of the water in the saturated gas stream is removed by a vane separator, a condenser, and a second vane separator. Finally, the off-gas is heated to a point where it becomes unsaturated. The gas is then filtered through two high efficiency particulate air filters and exhausted by an induced blower fan.

The system pictured in Figure 3.14 is constructed for testing radioactive soils. For soils containing organics, the dewatering and filtration of the gas stream may not be necessary (Buelt et al., 1985). However, a final treatment step using activated carbon or some other absorbent may be required.

In situ vitrification is a new technology that is not commercially available at this time. Although testing programs have provided promising results, much developmental work must be done to assess the applicability and cost effectiveness of this technique when applied to various waste types, particularly for the treatment of organic constituents. The limits of this process with regard to large-scale applications are also in need of further definition.

Disposal Needs

Depending on local regulations and off-gas chemical concentrations, disposal may be required for residuals from the off-gas treatment system (i.e., spent carbon, scrubber blowdown, etc).

Monitoring Requirements

Since few large-scale vitrification projects have been performed up to this time, it is unclear what the long-term monitoring requirements may be. The vitrified mass is expected to exhibit stability for a very long time. However, standard environmental monitoring (groundwater and soil) will be needed to provide information regarding the system's effectiveness.

The vitrification process requires a high degree of technical expertise to implement. Power requirements must be monitored to ensure an even melt, and the off-gas hood and treatment system require constant monitoring to ensure safe and adequate treatment of the off-gas stream.

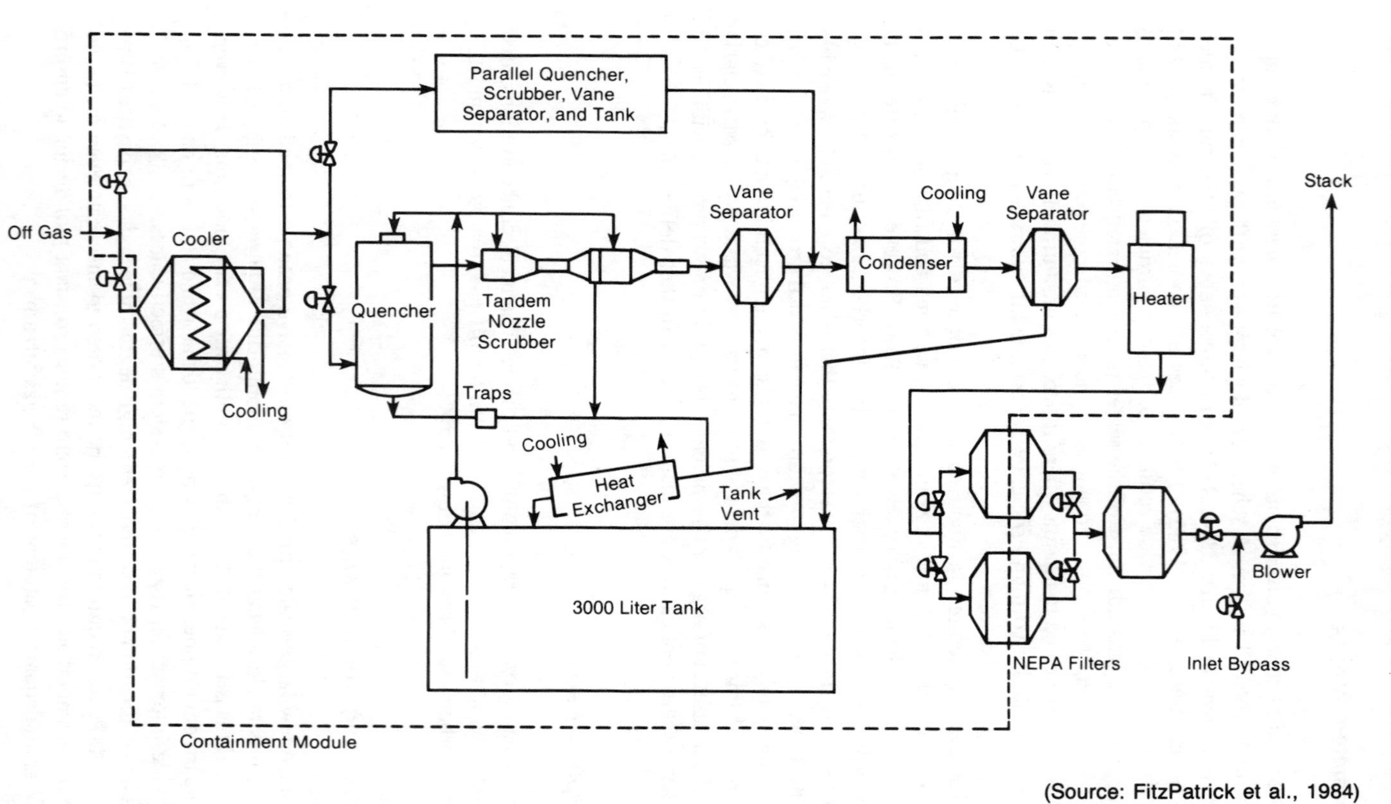

(Source: FitzPatrick et al., 1984)

Figure 3.14 Schematic for large-scale off-gas system.

Permitting Requirements

As with monitoring requirements, permitting requirements are unclear at this time. However, air discharge from the process may require an air discharge permit. Scrubber water discharge, if any, may require a National Pollution Discharge Elimination System (NPDES) permit or sewer discharge permit.

Environmental Feasibility of In Situ Vitrification

Exposure Pathways

During installation, operation, and monitoring of an in situ vitrification system, possible primary exposure pathways include the following:

- Inhalation of vapors from soil or groundwater during system installation and monitoring and/or vapor inhalation from the treatment unit during operations and monitoring.
- Dust inhalation and ingestion of particles during installation, monitoring, and treatment.
- Contact of skin with soils or groundwater during installation and monitoring and with operations, monitoring, and treatment.
- Primary impact to groundwater and surface water during installation, monitoring, treatment, and operation. This may determine the regulatory cleanup levels required if the waters have been affected.

Secondary pathways include dust inhalation and ingestion of particles during operation and monitoring when these impacts are minimized.

The consequences of human intrusion into a vitrified area can be significantly reduced by backfilling or covering the vitrified zone with some other sort of engineering barrier (Oma et al., 1983). Test results, such as the case study described, indicate that little compound migration occurs during the vitrification process.

Environmental Effectiveness

Organic materials appear to be combusted and/or destroyed by the high temperatures encountered during the vitrification process. This has been verified in the case study described earlier and also by tests on radioactive soils containing mixed wastes such as carbon tetrachloride, tributyl phosphate, dibutyl butylphosphate, wood, plastics, and other organic components (Buelt et al., 1985).

Long-term performance assessments (10,000 years) have been made to determine the effectiveness of selective vitrification for immobilizing highly concentrated zones of transuranic wastes (Oma et al., 1983). While the results of these performance assessments cannot be directly extrapolated to soils containing other materials, they do indicate that the in situ vitrification technology has the potential to isolate organic and inorganic materials.

Large-scale vitrification units are designed to operate to depths of 9.1 to 12.2 meters (30 to 40 ft). Material buried below this zone will be protected from leaching by the vitrified cover, but may be subject to groundwater contact in areas with high or fluctuating watertables.

Economic Feasibility of In Situ Vitrification*

Capital Costs

The cost of using a large-scale in situ vitrification technique to vitrify a typical 90×30×5 meter (295.2×98.4×16.4 foot) site has been estimated (PNL, 1986). Table 3.11 includes estimates of expenses for the following categories: 1) site preparation and closure costs; 2) annualized capital and site specific costs; 3) labor and operating costs; and 4) electrode and power (consumable) costs (PNL, 1986). Total estimated costs for large-scale in situ vitrification settings are expected to range from $150.3 to $222.2 per cubic meter ($115 to $170 per cubic yd^3). Site and equipment costs account for approximately 18 percent of the total annual cost.

Installation Costs

Labor costs for both a heavy equipment installation crew and the vitrification crew account for approximately 24 percent of the estimated total annual cost.

Operation and Maintenance Costs

Consumable costs associated with the operation of the vitrification system account for approximately 60 percent of the total cost for a large-scale vitrifi-

*All costs are presented in 1986 U.S. dollars unless otherwise specified.

cation project. Total cost of the project is significantly affected by electrical costs, as shown in Table 3.11.

Qualitative Ranking of Cost

The anticipated cost of in situ vitrification appears to be relatively high compared with other remedial actions for petroleum-laden soils. Costs are expected to decrease as the technology becomes more readily available.

IN SITU PASSIVE REMEDIATION

General Description

In situ passive remediation involves no action at the site for soil or groundwater. This type of remediation relies upon several natural processes to destroy the compounds of interest. These natural processes include biodegradation, volatilization, photolysis, leaching, and adsorption. These processes as well as the feasibility of this method, are described in detail in this subsection.

In situ passive remediation is by far the least expensive of all cleanup actions; however, it is generally unacceptable to the regulatory agencies.

Technical and Process Description

Passive remediation relies on various natural processes which can assist or possibly accomplish site remediation if allowed to proceed without external interferences. These processes depend on a variety of site- and chemical-specific parameters which are described briefly below:

- Biodegradation—Soil is the natural habitat for large numbers of microorganisms many of which can convert petroleum products to carbon dioxide and water. In situ microbial species that utilize the petroleum product as a carbon source will tend to thrive in the area of the underground spill and multiply preferentially, thereby enhancing biodegradation. (A more detailed explanation of biodegradation processes is provided in the subsection, In Situ Biodegradation.)
- Adsorption—Hydrocarbons in the subsurface may adhere to soil particles and resist further downward migration. This is particularly true when the hydrocarbons are of relatively high molecular weight and low water solubility. Adsorbed hydrocarbons will remain bound until acted upon

Table 3.11 In Situ Vitrification Cost Estimate (1985 $)

Cost Breakdown	Line Power $0.029/kWh[a]	Line Power $0.049/kWh[a]	Portable Power $0.0025/kWh[a]	Specific Basis
Site Costs				
Site Costs	25,000	25,000	25,000	Access, leveling, etc.
Site Costs/m³	2	2	2	Original soil volume basis
Equipment Costs				
Design and Engineering	468,000	468,000	715,000	25 percent of equipment costs
Portable Generators	76,000	76,000	1,090,000	
Power Line	26,000	26,000		
Transformer	213,000	213,000	213,000	
Electrode Power Cables	14,000	14,000	14,000	
Electrode Frame and Hood	129,000	129,000	129,000	
Electrode Placement Machinery	88,000	88,000	88,000	
Crane	96,000	96,000	96,000	
Front End Loader	64,000	64,000	64,000	
Off-Gas Equipment	1,169,000	1,169,000	1,169,000	
Total Equipment Cost	2,342,000	2,342,000	3,577,000	
Annualized Fixed Charge Rate	0.200	0.200	0.200	
Annualized Equipment Charges	468,000	468,000	715,000	
Annual Vitrification Rate (m³)	17,900	17,900	17,900	80 percent operating capacity
Site Vitrification Settings	48	48	48	
Site Volume (m³)	13,500	13,500	13,500	90 × 30 × 5 m site
Site Equipment Costs	353,000	353,000	540,000	
Equipment Cost/m³	26	26	40	Original soil volume basis

Labor Costs				
Vitrification Crew				
Operator	275,000	275,000	275,000	2 operators/shift @ $25/hr
Maintenance	13,000	13,000	13,000	1 hr/day @ $48/hr
Engineer	110,000	110,000	110,000	8 hr/day @ $50/hr (1 shift coverage)
Heavy Equipment Crew				
Operator	36,000	36,000	36,000	19 hr/setting @ $39/hr
Laborer	60,000	60,000	60,000	34 hr/setting @ $37/hr
Electrician	9,000	9,000	9,000	4 hr/setting @ $45/hr
Total Labor	503,000	503,000	503,000	
Labor Cost/m³	37	37	37	Original soil volume basis
Consumable Costs				
Electrodes	742,000	742,000	742,000	4 electrodes/setting, $770/m of electrode
Energy Consumption/Setting (kWh)	305,000	305,000	305,000	
Energy Cost	425,000	717,000	1,208,000	Cost for all settings
Total Consumables	1,167,000	1,459,000	1,950,000	
Consumable Cost/m³	86	108	144	Original soil volume basis
Total Site Cost	2,048,000	2,341,000	3,017,000	
Total Cost/m³	152	173	224	Original soil volume basis
Total Cost/cu ft	4.30	4.91	6.33	Original soil volume basis

Source: PNL, 1986

[a]This estimate reflects the expected operational costs of the In Situ Vitrification process applied to PCB-contaminated soils. The estimate should in no way be construed to reflect an official quotation by Battelle Northwest to provide equipment or technology to a potential user of the technology.

by other natural mechanisms such as leaching by infiltrating water.

- Volatilization—Certain hydrocarbons will migrate to the subsurface or surface and vaporize to the atmosphere. This is particularly true of constituents with a lower molecular weight and a high Henry's Law Constant. Detailed descriptions of remedial actions using enhanced volatilization are provided in the subsection, In Situ Volatilization.)
- Leaching—The flushing of hydrocarbons further downward within the unsaturated zone is known as leaching. Precipitation and surface infiltration are the two main driving forces and predominantly affect compounds with relatively low molecular weight and high water solubility.
- Photolysis—Sunlight can supply the activation energy required to transform hydrocarbons to other chemical compounds. This process is a surface phenomenon and has minimal impact on subsurface petroleum-laden soils (Brookman, et al., 1985).
- Soil permeability—Low soil permeability is favored for passive remediation. Clay and low permeability soils may form an aquitard which can minimize the effects of leaching and the ultimate impact to groundwater by leaching of water soluble components. On the other hand, greater permeability will enhance volatilization of lighter components and will also improve natural oxygen transport to the subsurface which will enhance biodegradation.
- Depth to groundwater—A shallow groundwater table will be more vulnerable to impact regardless of the petroleum product involved. Groundwater tables at greater depths allow natural adsorption and biodegradation mechanisms to minimize downward leaching of hydrocarbons.
- Infiltration—The rate of infiltration will determine the amount of water available for leaching and is dependent on the local amount of precipitation and soil permeabilities. Climates with very little rainfall and sites having low soil permeabilities will experience minimal infiltration and subsequent leaching. Extremely arid regions however may find depressed biological activity due to the lack of water for microbial growth and transport.
- Petroleum product composition—The chemical composition of the petroleum product is the single greatest influence on the operating mechanisms for passive remediation. Heavier, lower solubility, less volatile components will tend to remain bound to soil particles in the unsaturated zone. Leaching of these components over time tends to be minimal. Lighter petroleum products with more soluble components will be readily leached and could reach the water table. The volatility of these components may somewhat offset the impact of leaching. In situ microbial species favor lower molecular weight components and will utilize them preferentially though biodegradation of larger molecular weight components will occur over time.

A summary of the factors that influence the natural processes for passive remediation can be found in Table 3.12.

Technical Feasibility of Passive Remediation

Passive remediation, sometimes referred to as a "no-action alternative," is applicable to sites where the environmental conditions and regulatory policies are such that no other remedial action is required. Although environmental and waste characteristics are at times amenable to passive remediation, this alternative is rarely implemented because of local regulatory policies and site conditions.

Experience

A well-documented case of passive remediation is described by Odu (1972). During the period of mid-July to mid-August, an oil well blowout resulted

Table 3.12 Factors That Influence the Natural Processes for Passive Remediation

Soil Factors	Environmental Factors	Chemical Factors	Management Factors
Water Content	Temperature	Chemical Composition	Depth of Incorporation
Porosity/ Permeability	Wind	Concentration	
Clay Content	Evaporation		Irrigation Management
Adsorption Site Density	Precipitation Microbial		Soil Management
pH	Community		Availiability of Nutrients
Oxidation/ Reduction Potential			

Notes:

1. The factors listed above will not affect all natural processes in the same manner. For example, extremely high temperatures will enhance subsurface volatilization and inhibit biodegradation.
2. The effectiveness of passive remediation depends on complex relationships among the processes of volatilization, biodegradation, leaching, adsorption, and photolysis and is a function of all the factors listed above.

in a pool of crude oil 15.24 meters (50 ft) in diameter in the Ogoni Division of the Federal Republic of Nigeria. In this area soils are well drained sands or sandy loams, are moderately acidic (pH of 5), have approximately 2½ percent organic matter, have clay contents varying between 7 and 15 percent, and are deficient in nitrogen and potassium. A rainfall of approximately 2.54 meters (100 inches) per year is common for the area.

The total area affected by the crude oil flow and wind-carried spray was approximately 370,650 km² (1,500 acres), and four distinct regions were defined by extent of impact. The most heavily impacted region consisted of standing pools of oil, some knee-deep; the lightest impacted region initially showed only slight traces of oil. Approximately two weeks after the blowout was terminated, the total impacted area was reduced to 190,267 km² (770 acres). The predominant mechanisms were found to be leaching caused by rains and natural biodegradation.

Subsequent testing highlighted the following:

- Even in the most heavily contaminated areas significant amounts of oil were limited to the top 12.7 cm (5 inches) of soil. This was surprising in light of the sandy nature of the soils. Quantitative measurement of water soluble component migration was not done.
- Increases in microbial population were documented particularly in the areas of moderate to heavy oil concentration. Microbial population ranged from 2.6 million microbes/gm in moderately contaminated soils to 37.5 million microbes/gm in areas of heavy contamination.
- Tracking of microbial population showed initial depression of microbial numbers particularly in areas of heavy contamination. Recovery from this initial depression was quick. By the end of six weeks, increases in population indicated either general stimulation of natural microbial species or more likely preferential stimulation of species which utilized the petroleum compounds.

Although this study did not monitor groundwater for migration of hydrocarbons, a profile of the "ideal" environment, one that minimizes surface migration, is possible. Such an environment would be composed of a very shallow 0–15.24 cm (0–6 in.)) surface layer of moderately porous soils which would enhance volatilization and biodegradation, underlain by a very low permeability stratum which would act as a barrier to continued vertical migration.

Implementation Feasibility of In Situ Passive Remediation

Design Considerations

The decision to utilize passive remediation involves numerous site-specific as well as regulatory considerations. The factors favoring the possible suitability of this approach are:

- The regulatory policies allow consideration of passive remediation.
- The site is far from downstream receptors. Neighbors should be distant and on municipal water supplies.
- The depth to groundwater is large.
- A favorable soil profile is present. A permeable surface soil layer enhances biodegradation and low permeability soils between the surface soils and water table and inhibits downward migration of infiltrating water. Hydraulic conductivity ranges from greater than 10^2 cm/sec for highly permeable soils to less than 10^{-6} cm/sec for low permeability soils.
- The land is microbially fertile with populations of approximately 10 million microbes/gm or more.
- Rainfall is moderate to light with only a small probability of surface ponding.
- The spilled petroleum product is highly biodegradable and soil concentrations are initially low enough for biodegradation to proceed.

Equipment Requirements

Due to the nature of passive remediation, no equipment is required for the implementation or operation of the process. However, in order to prove the effectiveness of passive remediation at a site, an environmental monitoring system should be installed and maintained (see Monitoring Requirements).

Treatment Needs

Although the initial phases of passive remediation require no treatment, treatment may be needed in later phases to enhance biodegradation. Degradation of hydrocarbon components by natural biological processes is limited by various in situ factors as explained earlier. Additionally, the lowering of hydrocarbon concentration is in itself a limiting factor. It is generally agreed that below 1 mg/kg (ppm) the lack of a carbon source for utilization depresses further degradation. To restimulate this process, nutrients and oxygen can be delivered to the subsurface until the desired residual hydrocarbon concentration is reached. Heterotrophic microorganisms are the most common group of microorganisms providing the metabolic process for removing organic compounds from contaminated groundwater and soils. Heterotrophics use the same substances as sources of carbon and energy which can be obtained through three methods, fermentation, aerobic respiration, and anaerobic respiration (Canter and Knox, 1985). The subsection on In Situ Biodegradation describes in detail the treatment required for enhanced biodegradation.

Additional treatment needs may also arise based on monitoring results. Naturally occurring volatilization is limited as is biodegradation. Once equilibrium conditions are reached, volatilization will cease. Enhancement of this process is possible through in situ volatilization (discussed under In Situ Volatilization). Leaching to the groundwater may result in the need to pump and treat this stream (see Section 4, Groundwater Extraction and Treatment).

Disposal Needs

Unless additional treatment is required, passive remediation does not generate by-products or end-products requiring disposal. A potential end-product that could require disposal is the in-place soil if passive remediation is not successful.

Monitoring Requirements

Monitoring of the migration of petroleum compounds in the soil and water should be considered and will probably be required by regulatory agencies. Analysis of periodic soil and groundwater samples will document the effectiveness of this treatment as well as provide an early warning of possible impact to groundwater. Soil samples should be collected periodically in the affected area and analyzed for microbial population and concentration of petroleum compounds of concern. Groundwater monitoring wells should be installed downgradient of the petroleum source in an unimpacted region. At least one upgradient well should also be installed to ascertain, for comparison, the quality of groundwater entering the site. Depending on the nature of the leak or spill ambient air monitoring may also be advisable.

Permitting Requirements

Regulatory agencies may require verifiable data documenting the effectiveness of this treatment and environmental monitoring. Periodic soil and groundwater samples may be required. Depending on the composition of the petroleum product, permits for air emissions may or may not be required.

Passive remediation may be difficult to obtain. This is due to the fact that this is a "no-action" alternative. Regulatory agencies seldom agree that "no-action" is the appropriate remedial action. Extensive site characterization and demonstration studies may be required to show that adverse environmental and human impacts will not occur.

Environmental Feasibility of In Situ Passive Remediation

Exposure Pathways

Because passive remediation requires no excavation and little if any handling of the affected soils, the primary human exposure pathways during the duration of the remedial alternative are limited to inhalation of emissions from the surface soils. These pathways depend on the chemical and physical properties of the petroleum compounds and can be evaluated using models such as the SESOIL model described in Section 2.

However, because the choice of passive remediation requires initial site characterization and will most likely require ongoing monitoring, the pathways associated with these steps must be considered. The primary pathways associated with both site characterization and ongoing monitoring are related to soil and groundwater sample collection and include the following: inhalation of particles and volatilized hydrocarbon compounds and possible ingestion of soils or water containing the petroleum constituents. Section 2 describes these exposure pathways and applicable petroleum constituents in more detail.

Environmental Effectiveness

The effectiveness and suitability of passive remediation depend upon a variety of site-specific and constituent-specific factors. In addition to the site-specific factors mentioned under In Situ Volatilization, natural mechanisms such as biodegradation and volatilization are more effective when the hydrocarbons are composed of compounds of predominantly lower molecular weight. Although this favors both biodegradation and volatilization, compounds with lower molecular weights also favor leaching. Heavier components tend to remain bound in the soils. Again, because these factors are extremely site-specific, the environmental effectiveness of passive remediation will be decided on a case-by-case basis.

Economic Feasibility of In Situ Passive Remediation*

Capital Costs

Capital costs associated with passive remediation can be modest. Periodic soil sampling may require the purchase of hand-held core samplers, such as

*All costs are presented in 1986 U.S. dollars unless otherwise specified.

the Wilcox Corer which costs $100–$200 per sampler. Monitoring of groundwater quality will require monitoring wells that cost approximately $65.6 per linear meter ($20 per linear foot) for Schedule 80, 15.24 cm (6 in.) diameter PVC pipe.

Installation Costs

The installation cost of 6-inch PVC monitoring wells ranges from $49.20 to $65.60 per meter ($15 to $20 per foot) of depth.

Operation and Maintenance Costs

Periodic maintenance of wells will require little expense if installed correctly. Well maintenance expenses will be related to well rehabilitation if required to remove scaling, iron-bacteria, or other encrustations or siltations of the well screen. Costs for labor to obtain well samples and laboratory services to process samples will depend upon the need to purchase labor (vs using existing personnel) and the frequency and type of sampling required. Frequent sampling for an extensive list of constituents can cost over $10,000 annually.

Qualitative Ranking of Cost

Passive remediation is the least expensive treatment option of all options considered in this report.

IN SITU ISOLATION/CONTAINMENT

General Description

Isolation/containment is the process by which the area of concern is separated from the environment. This separation can be accomplished by the installation of containment devices, some of which are caps, cut-off walls, grout curtains, and slurry walls. This process attempts to isolate the site so that any compounds of concern are immobilized within. The containment devices, as well as the process feasibility, are discussed in detail in this subsection.

Experience has proven that the containment devices discussed in this section adequately isolate the contamination. However, destruction of the compounds is not accomplished.

Process Description

Methods have been developed to contain petroleum products in place within the unsaturated (vadose) zone. These methods minimize the migration of hydrocarbon compounds by:

- Subsurface physical barriers placed in the path of flow and typically used to control lateral flow which minimizes the spread of hydrocarbons.
- Surface covers or caps typically used to reduce surface infiltration from precipitation and surface runon.

Capping

The reduction of infiltration can be achieved through "capping" with impervious materials or through surface-sealing techniques. Many methods exist for capping. These can be generally grouped into the following classes:

- Synthetic membrane.
- Low permeability soils.
- Soil/bentonite admixtures.
- Constructed cap, asphalt, or concrete.
- Multilayered cover system.

Synthetic Membrane Caps

Several varieties of synthetic membranes may be applicable for use in capping sites to prevent infiltration of precipitation. In some applications, synthetic membranes may offer substantial cost benefits over the use of other options such as compacted clays or other low permeability soils and mixtures. Membrane materials include polymers, rubbers, coated fabrics, and others. Chemical compatibility between the synthetic membrane and the hydrocarbon must be considered as part of the selection process.

The successful use of synthetic membranes depends upon the selection of the proper membrane material for the desired application; proper installation including seaming and placement to prevent tearing; and protection against weathering, root penetration, or other surface damage. The major benefits of synthetic membranes are their availability and extremely-low permeabilities. The major design limitations of synthetic membranes is their potential for failure in the long term due to damage caused by puncturing, tearing, or weathering. These can be addressed through proper cap system design, installation, and surface protection.

Low Permeability Soils

The term "low permeability soils" refers to those fine-grained soils that when compacted, consistently maintain an in situ permeability of 10^{-6} to 10^{-7} cm/sec (0.1 ft/yr) or less. A key advantage in the use of compacted low permeability soils is that they are a natural material and are often more durable in the long term.

The compacted low permeability soil cover is constructed after the site is prepared and fill materials are placed. The cap is then covered by a clean soil layer followed by top soil and vegetation. The primary advantages of low permeability soils are that they are natural materials, should exhibit longevity, and are self-healing to some extent. A clay cap could be expensive, depending on availability, and will require surface protection to maintain its integrity (due to root penetration, freeze thaw, drying/cracking, etc.).

Soil/Bentonite Admixtures

A low permeability soil/bentonite admixture can be placed as the cap layer in a multilayer cap system described later, or as a single layer cap. Soil/bentonite admixtures are a proven capping technique in waste management and may be considered as a possible substitute for low permeability soil when it is not readily available.

The installation process typically incorporates a geotechnical assessment of the available soils for use in the admixture and a determination of suitability followed by a determination of the necessary bentonite application rate to achieve the desired cap effectiveness. The bentonite is placed and "admixed" with the soils. The mixture is then uniformly spread and compacted. The bentonite, once hydrated to the optimum moisture content, swells to fill the void spaces within the soil layer and an effective seal is achieved to control site infiltration.

In general, fine-grained soils have better applicability with bentonite admixtures since their pore size openings are smaller and their inherent permeability rates are less than coarse-grained soils. Clay soils, however, are an exception. The cohesive nature of clays, in general, makes admixing operations difficult. The more suitable soils tend to be inorganic soils and poorly-graded sand/silt mixtures.

Constructed Cap (Asphalt or Concrete)

An asphalt or concrete cap can be an effective means to control surface infiltration. These cap materials typically involve a higher cost than low permeability soils or synthetic membranes. However, low permeabilities can be achieved using the constructed cap and the paved area would be suitable for other light uses (i.e., parking lot, storage, etc.). The constructed cap should be sloped for drainage and will require periodic maintenance to repair cracking and weather damage. The subgrade must be prepared to provide a stable base and handle the expected loadings.

Multilayer Cap

The multilayer cap system represents a cover technology that is gaining widespread use in the field as an infiltration control strategy for waste containment or in-place closure. The multilayer cap system performs the basic functions of minimizing infiltration; directing and transmitting percolation away from the site; and proving a final cover for the site and growth medium for vegetation. The multilayer design is a more efficient system than the standard clay cap in diverting infiltration from the underlying soils. A typical multilayer cap system, as shown in Figure 3.15, consists of the following three layers:

- *Upper soil layer.* A top soil and native soil layer, typically placed to a depth of 0.3 to 0.6 meters (12 to 24 in.). This layer serves to support vegetation, provide a cover for the drain layer, and divert surface runoff.
- *Middle drain layer.* A graded layer of porous flow zone material (e.g., sand, geogrid) to act as a drainage medium. This layer is typically placed to a depth of about 0.46 meters (18 in.).
- *Cap layer.* A compacted layer of fine-grained soils of low permeability designed to divert infiltration that has percolated through the upper soil layer. This cap layer is typically placed to depths of 0.46 to 0.61 meters (18 to 24 in.) and is the bottom layer of a multilayer cover system.

Several major advantages of the multilayer cover system, as compared to a standard native soil cover, include the following:

- A protective soil layer is placed over the cap layer; the cap is not directly exposed to potential damage due to weathering, cracking, or excessive root penetration.
- A drain layer serves to divert additional percolating water so it does not eventually migrate through the cap and into the underlying waste material.
- Possible slumping of the topsoil and upper soil layers is minimized, particularly in slope areas.

Groundwater Containment

Groundwater containment technologies, which can be passive or active, are

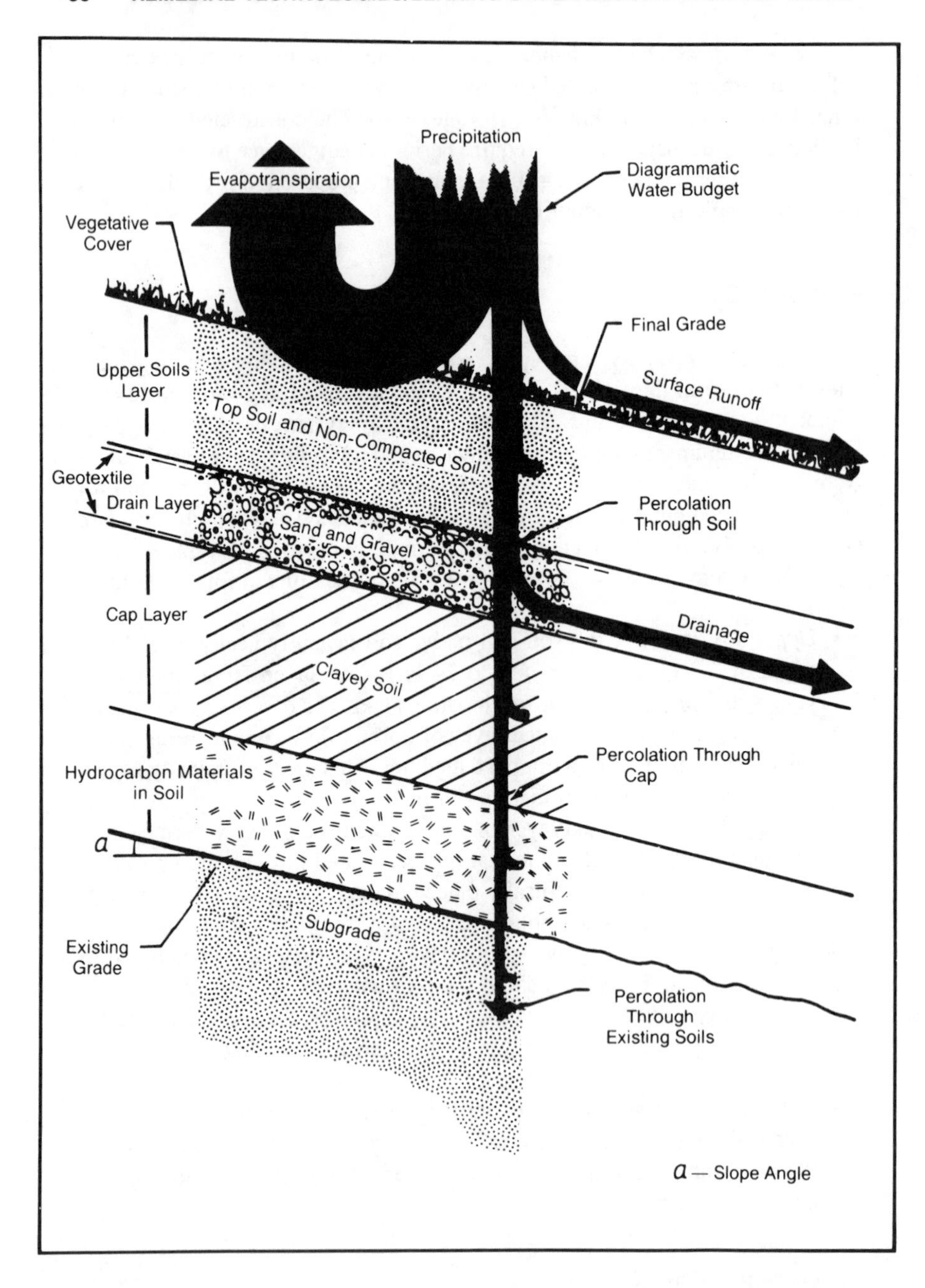

Figure 3.15 Typical multilayer cap system profile.

designed to control the lateral flow of groundwater and contain a hydrocarbon plume. The primary objective of these measures is to redirect the flow of groundwater around a spill site or contain affected groundwater to a finite region. The successful use of these measures depends on the specific hydrologic and soil conditions of the site.

Methods used in passive groundwater diversion technologies are outlined in this subsection. These methods include slurry cutoff walls, grouting techniques, and sheet piling, all of which require materials compatible with native soils and the waste streams. In contrast, active groundwater diversion involves actual pumping and subsequent groundwater treatment and is detailed in Section 4, Groundwater Extraction and Treatment.

Slurry Walls/Cutoff Walls

Slurry walls are fixed underground physical barriers formed by pumping slurry, usually a cement or bentonite and water mixture, into a trench as excavation proceeds, and either allowing the slurry to set (for cement-bentonite slurries) or back filling with a suitable engineered material (for soil-bentonite slurries). Typically, the maximum depth of a slurry wall is limited to the excavation reach of a large backhoe. Specialized equipment can achieve greater depths. The slurry itself is used primarily to maintain the trench during excavation. The weight of the slurry forces bentonite to penetrate the voids in the soil matrix. The effect of this is to cause the sides and the bottoms of the trench to be lined with a layer of bentonite thereby creating a wall of reduced permeability. Figure 3.16 depicts a typical slurry wall. In order to be effective, a slurry wall can be connected (keyed) to a low permeability stratum such as an aquiclude or to a competent geological member or be constructed "hanging." A hanging slurry wall may also be effective; however, soluble constituents may pass under the wall with the groundwater flow. A hanging slurry wall will effectively contain floating contaminants. Additional remedial groundwater extraction and treatment techniques are typically used in conjunction with a subsurface barrier to remove the contained groundwater for treatment.

Grout Curtain Technique

Grout curtains are fixed underground physical barriers formed by injecting grout, either particulate (such as portland cement) or chemical (such as sodium silicate) into the ground through well points. Grout curtains are typically designed to perform the same function as a slurry cutoff wall. Figure 3.17 depicts the feature of a grout curtain in combination with a groundwater recovery well. This grout curtain technique will accomplish the following:

Keyed-In Slurry Wall

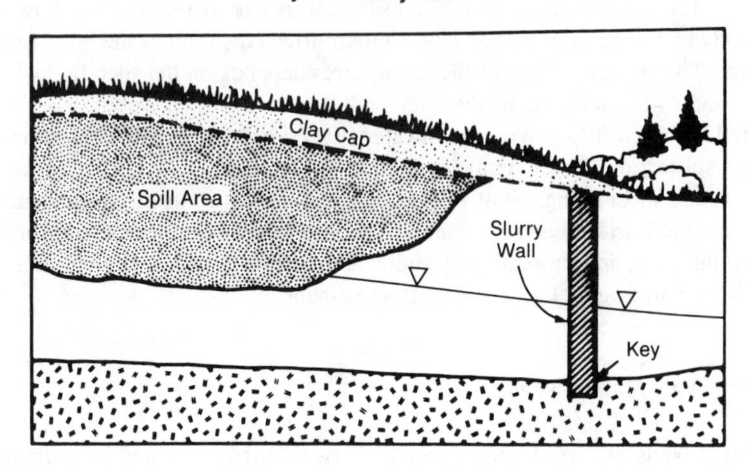

Hanging Slurry Wall

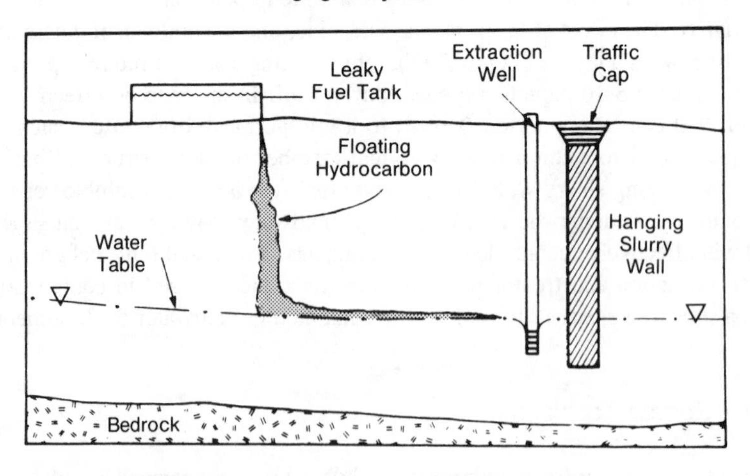

(From "Slurry Trench Construction for Pollution Migration Control," EPA-540/2-84-001, 1984)

Figure 3.16 Keyed-in slurry wall/hanging slurry wall.

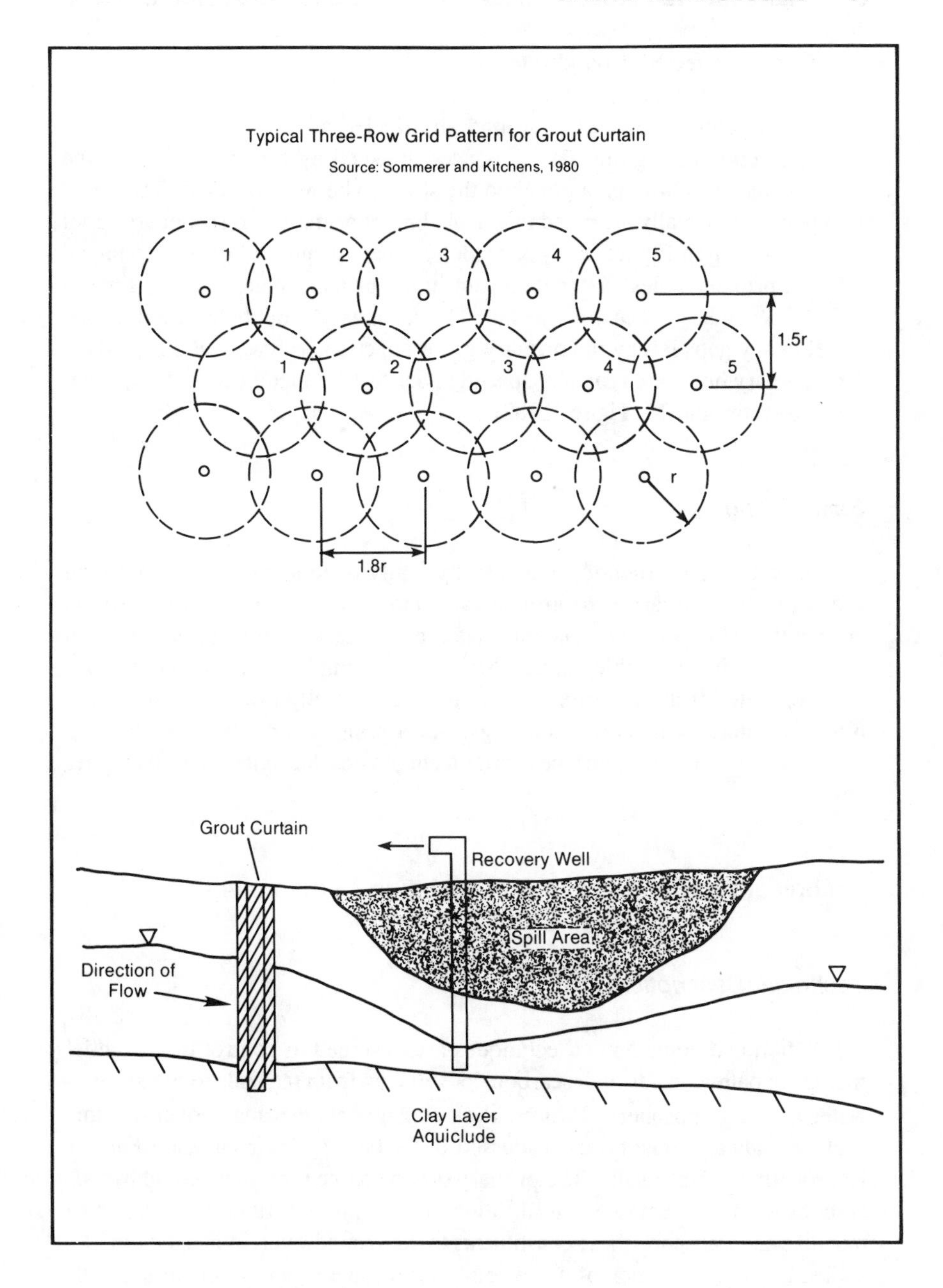

Figure 3.17 Upgradient grout curtain.

- Contain affected groundwater.
- Divert a chemical plume.
- Divert groundwater flow around an affected area.

Construction of a grout barrier is accomplished by pressure injecting the grouting material through a pipe into the strata to be waterproofed. The injection points are usually arranged in a triple line of primary and secondary grout holes as shown in Figure 3.17. A predetermined quantity of grout is pumped into the primary holes. After the grout in the primary holes has had time to gel, the secondary holes are injected. The secondary grout holes are intended to fill in any gaps left by the primary grout injection (Hayward Baker, 1980). The primary holes are typically spaced at 6.1 to 12.2 meter (20 to 40 ft) intervals (Guertin and McTigue, 1982).

Sheet Piling

Sheet piling construction involves physically driving rigid sheets into the ground to form a barrier to groundwater movement. Typically these sheets are composed of steel or concrete that can be interlocked or sealed to form a continuous impermeable barrier. Steel sheet piling is more commonly used than concrete for groundwater cutoff due to a generally lower cost and capability for interlocking between pilings. Sheet piling has not been used as extensively as the other subsurface barrier technologies due to its generally higher cost.

Technical Feasibility of In Situ Isolation/Containment

Technical Description

Isolation and containment techniques are designed to control the potential migration pathways for hydrocarbon constituents from the soils to the groundwater and/or atmosphere. Hydrocarbon constituents that may adversely impact groundwater quality are discussed in Section 2, Environmental Fate of Hydrocarbon Constituents. The primary driving force for movement of hydrocarbons to the water table is infiltration which causes certain constituents to mobilize and transports these constituents to the groundwater. When the product spilled has a high fraction of water soluble constituents, and when on-site soils are permeable and precipitation is plentiful, infiltration is enhanced. In contrast, the primary driving force for movement of hydrocarbons from soils into the atmosphere is volatilization. Placing a cover over affected soils to minimize the amount of water flushing through the soils and to reduce surface area will minimize infiltration and volatilization.

In addition to surface infiltration, the lateral movement of shallow groundwater can transport constituents within an aquifer. Slurry walls and similar subsurface barriers can be used to limit this lateral flow of groundwater.

Experience

Isolation and containment techniques have been used for many years to mitigate the effects of spills, cover existing disposal areas, and isolate a variety of containment areas. The case study presented below demonstrates the applicable use of capping, slurry walls, and pumping and treating of groundwater. Although this case did not specifically involve a spill of petroleum products, the same principles would apply.

Case Study—Slurry Wall and Capping. The Gilson Road hazardous waste disposal site is located in the City of Nashua, New Hampshire, in the southeasterly corner of that community. (See Figure 3.18.) The 1,482.6 km^2 (6-acres) site had been used as a sand borrow pit for an undetermined number of years. At some time during the late 1960s, after much of the sand had been removed from the property, the operator of the pit began an unapproved waste disposal operation (McGarry and Lamarre, 1985).

In 1980, sampling of newly installed groundwater monitoring wells revealed substantial concentrations of volatile organics, metals and extractable organics in the groundwater. The plume had extended over 4,942 km^2 (20 acres), contaminated a small surface stream, and was moving toward the Nashua River. Groundwater flow through the site was estimated to range from 49,140 to 245,700 liters (13,000 to 65,000 gal) per day depending upon the season of the year and groundwater gradients.

A slurry wall and surface cap system was designed to act as a clean water exclusion system at the site. The wall and the cap were designed to minimize the amount of unaffected groundwater entering the contained site. This system was designed to be used in conjunction with a groundwater recovery treatment process. Extraction wells would pump affected water which would then be treated and returned as recharge to the aquifer.

During the fall of 1982, a 0.91 m (3 ft) wide bentonite slurry wall was installed around the site along with a surface cap. The slurry wall totaled nearly 20,004.8 m^3 (26,150 yd^3) in cross section with a maximum depth of 32.9 m (36 yd) and surrounded a plume with a volume which was greater than 73,210.5 m^3 (95,700 yd^3). Excavation was accomplished using a modified Koehring backhoe capable of trenching to depths of 20.1 m (22 yds) in combination with

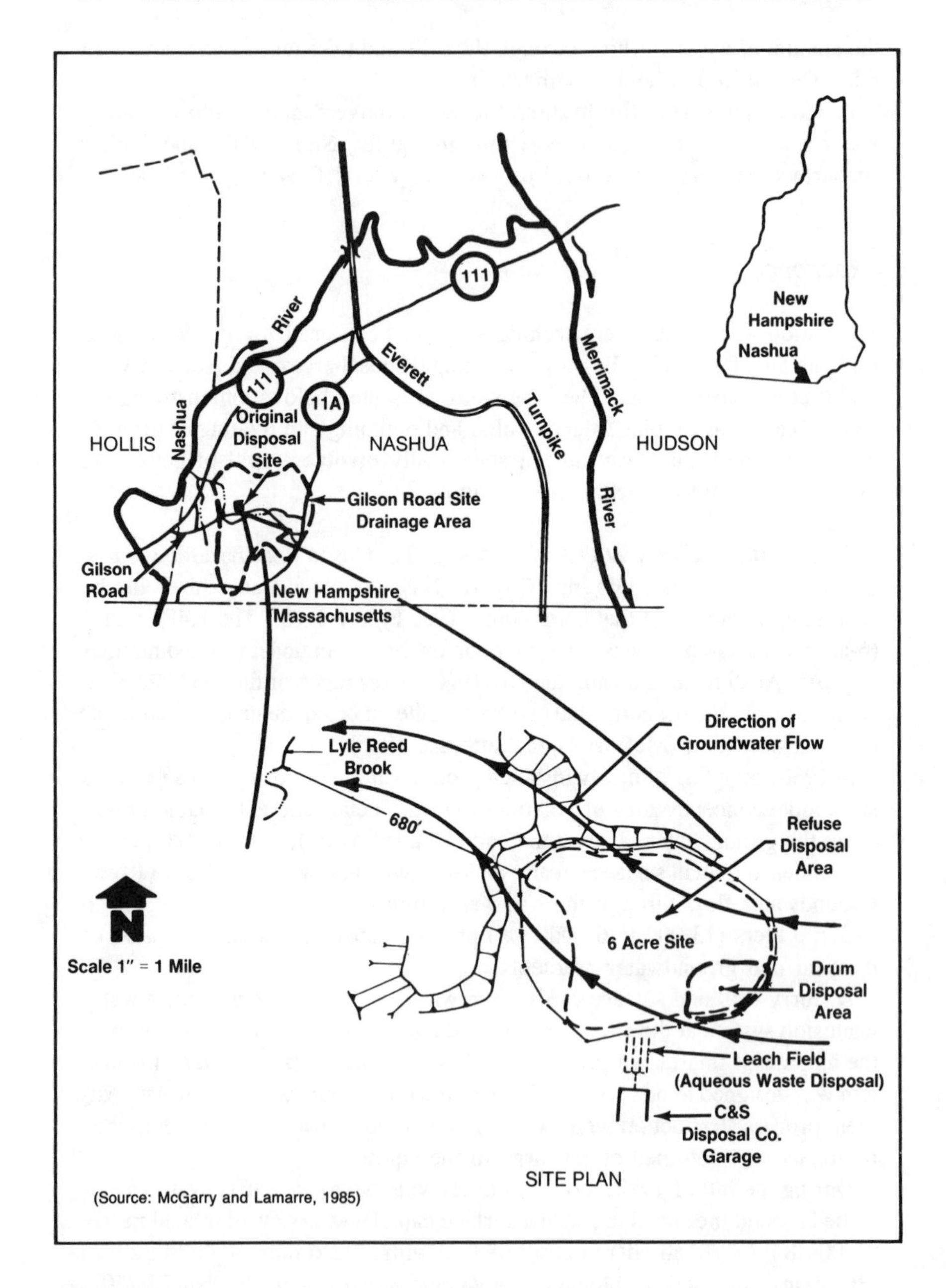

Figure 3.18 Gilson Road disposal site.

a 10.9 metric ton (12-ton) clam shell for depths greater than 22 yards. The one-meter wide trench excavation was keyed into bedrock and backfilled with a soil/bentonite slurry mixture. The entire excavation and backfilling process was completed in eight weeks (Ayers et al., 1983).

The cap consisted of a 60 mm high density polyethylene cover over which 0.3 to 0.76 meters (1 to 2.5 ft) of soil were compacted. No underdrains were provided for the cap.

Construction of a 1,134 Lpm (300 gpm) groundwater extraction and treatment system was also completed in 1985 and operation of the vadose zone commenced in January 1986. It was expected that two full flushes of the vadose zone over a period of two years would remove 90 percent of the compounds retained by the soil. At present, the project is on schedule, working as intended and achieving the expected 90 percent removal efficiency.

Implementation Feasibility of Isolation/Containment

Design Considerations

The design of any containment system must address all possible migration pathways. In the case of a spill of petroleum products, the containment system must, at a minimum, minimize percolation of precipitation through the saturated soils. The following design considerations are necessary and are common to most containment/isolation systems (Ehrenfelder and Bass, 1983; Moore, 1980):

- Divert surface water runon by utilizing dikes, diversion channels, berms, and other physical barriers to divert surface water around the spill area.
- Address site-specific conditions and the potential migration pathways.
- Reduce infiltration by using one of a number of capping alternatives that can reduce infiltration.
- Protect the cap and surface drainage controls from erosional damage. The use of appropriate grading for each construction material and a layer of topsoil and vegetation are often used to protect caps from erosion.
- Control lateral groundwater flow through the saturated soils with the use of slurry walls or other subsurface containment techniques. Slurry walls, however, must be compatible with the subsurface environment as well as the chemical compounds at the site.
- During the design and construction phases of the isolation/containment system, consider future use and use restrictions of the land. For example, future use of a capped area must not damage the integrity of the cap system, and site use restrictions should be in place.

Equipment Requirements

Generally, large earth-moving equipment is required for installation of slurry walls and caps. Large dozers, backhoes and a variety of earth-hauling trucks and machines are acceptable. Vibrating rollers for compaction are often used in addition to tracked and nontracked vehicles.

Treatment Needs

To enhance containment, isolation and containment techniques are typically used in conjunction with pump and treat systems which lower the water table and remove affected groundwater if present. Such systems pump the groundwater from inside the contained area for treatment. Depending on the nature of the petroleum product, a variety of treatment schemes are available including carbon adsorption and air stripping (see Section 4, Groundwater Extraction and Treatment).

Disposal Needs

Isolation and in situ containment preclude the need for disposal options. The use of pump and treat options in combination with isolation/containment may result in treatment and discharge requirements for the water stream along with disposal of treatment residues such as activated carbon.

Monitoring Requirements

The ultimate goal of the containment/isolation option is to stop the subsurface spread of petroleum products following a release. The containment/isolation technologies involve leaving hydrocarbons in-place. As a result, monitoring becomes important as a means for measuring effectiveness of the containment. Typically, this is accomplished by means of visual inspections of caps for structural integrity and groundwater monitoring to measure water quality and piezometric surface. Long-term monitoring may be required to ensure proper maintenance and longevity of containment.

Permitting Requirements

The use of pump and treat options associated with isolation/containment may require compliance with the National Pollution Discharge Elimination System (NPDES) for discharge of treated waters.

Environmental Feasibility of Isolation/Containment

Exposure Pathways

The purpose of isolation/containment is to prevent migration of volatile organic compounds within the vadose zone and groundwater and above ground surface. As a result, exposure pathways should be minimized.

Possible direct human exposure pathways during installation, monitoring, and operations include vapor inhalation and skin contact. Particle inhalation and ingestion are also considered primary human exposure pathways during installation.

Primary environmental exposure pathways which must be considered include impacts to groundwater and surface water.

Environmental Effectiveness

The effectiveness of a multilayer cap system can be evaluated using the HELP model (Hydrologic Evaluation of Landfill Performance). HELP is a computer model capable of predicting cap system performance (Barnes et al., 1986; Peters et al., undated). The following multilayer cap system was developed and used for the model simulation:

- Soil cover layer: 45.7 cm (18 in.) thick, vegetated.
- Middle drain layer: 30.5 cm (12 in.) thick, 4 percent slope.
- Cap layer: 45.7 cm (18 in.) thick, hydraulic conductivity of 0.00036 cm/hr (0.000142 in./hr).

Using 1974–1978 weather data for west-central Pennsylvania, the performance of the capping was simulated. The results of the simulation show that the multilayered cap system is effective in preventing more than 96 percent of the incident precipitation from reaching the underlying soil. The model also clearly shows the effectiveness of the lateral drainage layer, which allows more than 52 percent of the precipitation to drain off the cover layer and only 4 percent to reach and percolate through the low-permeability barrier layer. The model indicates negligible runoff with nearly all incident precipitation infiltrating the cover and with the largest percentage of infiltration being lost to evapotranspiration and drainage from the lateral drainage layer. Under field conditions runoff may contribute even a greater percentage to the reduction in infiltration.

Economic Feasibility of Isolation/Containment*

The long-term integrity of some of the materials used in isolation/containment structures has not been fully demonstrated. For example, there is some concern that clays and/or man-made membrane materials may not be compatible with certain organic constituents. This is an important factor in determining the life span of a slurry wall or a grout curtain. In addition, the ultimate life span of many synthetic materials used in the construction of cover systems is currently unknown.

Installation Costs

The cost of materials (EPA, 1985) and installation vary with type and extent of capping required. Generally, \$2.69 to \$21.50/m^2 installed can be expected for synthetic membranes with the variation due to thickness and chemical resistivity. The cost of purchase and placement of soils will vary from \$3.92 to \$13.07/m^3 (\$3 to \$10/yd^3) depending on clay content, distance to site, quantity ordered, and other factors. Bentonite mixtures will cost approximately \$10.75 to \$21.50/m^2 (\$1 to \$2/ft^2) of area covered, including placement. Generally, large earth-moving equipment is rented or a subcontractor is hired.

The cost for slurry wall installation varies with depth thickness and the use of soil or cement bentonite admixture (EPA Handbook, 1984). For a depth of less than 9.14 meters (30 ft) in a hard soil, the cost for a soil bentonite slurry wall could be approximately \$53.76 to \$107.53 m^2 (\$5 to \$10 ft^2) area covered. For a similar cement bentonite wall, the cost could be approximately \$268.82 to \$537.63/m^2 (\$25 to \$50/ft^2).

Maintenance and Monitoring Costs

Maintenance costs are generally modest for these systems. Monitoring for signs of erosion or settlement is recommended and generally occurs, if at all, in the first six months. Maintaining the integrity of a cap is of prime importance for a containment system. Trouble areas should be uncovered, repaired and recovered. Regular monitoring of groundwater around the perimeter of the area will likely be required for groundwater quality and the piezometric surface.

---

*All costs are presented in 1986 U.S. dollars unless otherwise specified.

Qualitative Ranking of Cost

Isolation and containment options are generally low in relative cost. The cost for excavation and treatment of soils is typically much more expensive. Monitoring costs and associated lab fees will add to the cost of this option; however, overall costs are still considered relatively low.

SECTION 4

NON-IN SITU TECHNOLOGIES

INTRODUCTION

The remedial technologies discussed in this section are non-in situ. Rather than accomplishing remediation in place, the non-in situ techniques require the removal, usually by excavation, of petroleum-laden soils. These soils can then be either treated on-site or hauled off-site and treated. In contrast to the in situ technologies, the non-in situ technologies must consider exposure pathways associated with the handling and/or transport of contaminated material.

The non-in situ technologies for soils discussed in this section include: Land Treatment, Thermal Treatment (low temperature and incineration), Asphalt Incorporation, Solidification/Stabilization, Chemical Extraction and Excavation. Technologies for treatment of affected groundwater also are discussed.

LAND TREATMENT TECHNOLOGY

General Description

Land treatment is the process by which affected soils are removed and spread over an area to enhance naturally-occurring processes. These natural processes include volatilization, aeration, biodegradation, and photolysis. How this is accomplished, as well as the feasibility of land treatment, is discussed in detail in this subsection.

Experience has proven that landfarming, if properly performed, is an effective method for the removal of hydrocarbons from affected soils. However, a great deal of available land and time can be required to accomplish hydrocarbon destruction.

Process Description

The land treatment or landfarming process involves the tilling and cultivating of soils to enhance the biological degradation of hydrocarbon compounds. Note that the 1.52 meter (5 ft) deep treatment zone includes a zone of incorporation (0.15–0.3 meter) (0.5–1 ft) near the surface where most degradation occurs and a deeper zone (0.3–1.52 m) (1–5 ft) where leachable components become immobilized and degrade more slowly. When leakage from an underground tank affects soils below or near the five-foot treatment depth, the soils must be excavated and reapplied at a surface site. When a surface spill affects shallow soils, excavation may not be necessary and landfarming "in place" may be possible.

The basic landfarm operations are as follows:

- The area which will be used for landfarming is prepared by removing surface debris, large rocks and brush.
- The area is graded to provide positive drainage and surrounded by a soil berm to contain run-off within the landfarm area.
- If necessary, the pH of the soil is adjusted with lime to provide a neutral pH. At neutral pH, metals are not highly mobile and bacterial processes and growth are supported.
- Agricultural fertilizer is added if the site is deficient in nutrients such as nitrogen, phosphorus, potassium or trace elements. Fertilizer is added as needed during the biodegradation process.
- The soils containing petroleum products are spread uniformly over the surface of the prepared area. It is important to distribute the hydrocarbons over the landfarm area as uniformly as practical to minimize localized loadings. Generally, petroleum products can be applied in quantities up to 5 percent by weight of the soil.
- The landfarmed material is incorporated into the top 15.24 to 20.32 cm (6 to 8 in.) of soil with a tiller, disc harrow, or other ploughing device. The soil must be well mixed to increase contact between the organics and microorganisms and to supply air for aerobic biological degradation.
- Depending on the rate of degradation, soils which contain petroleum products can be applied to the site at regular intervals. Reapplication at proper intervals replenishes the hydrocarbon supply and maintains biological activity.
- Monitoring of soils and surface runoff is typically conducted to measure hydrocarbon and nutrient levels and soil pH and to assure that the hydrocarbons are properly contained and treated in the landfarm area. Groundwater monitoring is also performed for large landfarming operations, and is required in many states.

Composting is similar to landfarming, because both processes rely on the destruction of organic compounds through microbial metabolism. Composting is a proven technology for achieving accelerated biodegradation of select industrial and municipal wastes under controlled conditions. Theoretically, all organic carbon waste is treatable using composting. Industrial, agricultural, and municipal wastes are being successfully treated using the composting process. Approximately 115 sludge composting facilities are currently operational in the United States (EPA, 1985).

Three general categories of compost systems are used: windrow, static pile, and in-vessel. In the windrow method, the mixture to be composted is piled in long rows (windrows) that are turned periodically by mechanical means to increase exposure of organic matter to oxygen. The static pile (forced-aeration) approach uses a blower to aerate the mixture to be composted. This mixture is placed upon a base of wood chips or other suitable material in which a network of aeration pipe has been constructed. Oxygen is then introduced by blowing or drawing air through the pile. In-vessel composting (mechanical or enclosed reactor composting) occurs in enclosed containers where environmental conditions can be controlled.

The majority of the composting systems in use are for treating municipal sludge. Approximately 90 percent of the operational facilities in the U.S. use static pile composting. Most of the remaining systems use windrows. In-vessel composting is currently being developed, and several systems are currently under design or construction.

The process flow is similar for all three composting sytems. The material to be composted is mixed with a bulking agent or agents such as wood chips, straw, horse manure, sawdust, leaves, or paper. The bulking agent can serve as a source of carbon, nutrients, or microbes. In addition, it increases porosity and aeration, as well as dilutes the concentration of the contaminant. This latter factor is important for composting materials such as explosives-contaminated soil. Once the mixture to be composted is in place, it undergoes a self-heating process caused by microbial activity.

After composting, the material to be composted is usually cured for about 30 days. During this period, additional decomposition, as well as stabilization, pathogen destruction, and degassing take place. Curing can be followed by a drying stage if the compost is to be screened for recovery of bulking agent.

The composting of petroleum-laden soils involves modification of normal composting systems and is described below (Perlin and Gilardi, 1978):

- Bulking agents such as wood chips, sawdust, straw, etc., are placed in strips about 0.38 meters (15 in.) deep by 4.6 meters (15 ft) wide on the ground. In order to keep all composting materials off the surface soils and prevent soil contamination, it may be desirable to pave the site.

- Petroleum-laden soils are spread over the bulking agents and then the two components are mixed together using a front end loader or windrow composter. Proper blending of the bulking agent and soil requires two or three mixings to ensure that homogeneity is achieved.
- The mixture to be composted is then configured according to the system to be used.
- Aeration is provided by natural convection currents and pile turning for windrows. Forced aeration is utilized for static pile or in-vessel systems. To ensure even degradation and proper aeration throughout the mass, windrows are remixed at regulated intervals, approximately 2 times a week. More frequent mixing may be required initially.
- After degradation plateaus (3 to 4 weeks), windrows are flattened and static piles distributed and left to dry to 30 percent moisture content. It is possible to add more soils to the compost as long as favorable conditions, such as proper aeration and moisture content, are maintained. Care should be taken not to add too much soil since anaerobic conditions could result from an inadequate supply of bulking agent. If conditions are too dry for degradation, water can be applied.
- Curing is the next stage. The composted mass may be stockpiled and stored for the desired amount of time, usually about one month. Biodegradation still occurs, but at a much slower rate than during the previous steps.

Composting is an attractive pretreatment method for landfarming. Because composting does not fully degrade all petroleum hydrocarbons, it is often followed by landfarming. Composting processes partially biodegrade some petroleum products. When partially degraded, these products are mixed with topsoil and render a more suitable environment for landfarming. In addition, the leaching potential of metals present in petroleum-laden soils may be reduced by composting.

Technical Feasibility of Land Treatment Technology

Technical Description

Land treatment and composting remove petroleum hydrocarbons from soils using a combination of the following processes: volatilization, leaching, incorporation of the hydrocarbons into the soil matrix by sorption, and degradation. Although volatilization removes a large portion of the lighter hydrocarbons, the effects of the first three of these mechanisms on the land treatment of heavy hydrocarbons are minor when compared to degradation. The various processes involved in landfarming are schematically illustrated in Figure 4.1.

In land treatment, degradation of petroleum hydrocarbons in soils is responsible for the decomposition of the heavier fraction of hydrocarbons and can occur chemically, photochemically, and/or microbially:

- *Microbial* degradation (or biodegradation) is the primary mechanism for decomposition of petroleum products. Biodegradation is the process by which microorganisms present in soil acquire energy and cellular nutrients through the breakdown of various petroleum hydrocarbons. This process takes place on a microscopic level and is the basis for landfarming. This process is described in detail in Section 3, In Situ Biodegradation.
- *Chemical* degradation occurs if conditions are favorable for a particular reaction, such as hydrolysis, oxidation, or reduction. Factors affecting chemical degradation include soil pH, redox potential, availability of catalytic sites, etc. When compared to microbial degradation, chemical processes perform a minor role in the decomposition of petroleum hydrocarbons.
- *Photochemical* degradation occurs when sunlight energy breaks the chemical bonds of organic constituents. Whether or not this occurs, depends on the quantity of sunlight to which the hydrocarbons are exposed. The photochemical contribution to degradation in land treatment systems is minimal.

Soil is the natural environment for large numbers of microorganisms; consequently, biodegradation can occur quite easily if physical and chemical conditions are appropriate and the substrate is biodegradable. Soil microbes are generally aerobic, especially in well-drained soils. Phung et al. (1978) state that aerobic microbial metabolism of hydrocarbons converts organics to carbon dioxide, microbial cells, and water as shown below:

AEROBIC BIODEGRADATION

Petroleum
Products
$(CHO)_nNS$ --------→ CO_2 + Microbial Cells + NH_4 + H_2S + Energy + H_2O

(60% of Original Carbon in Petroleum Hydrocarbons)

$NH_4 \rightarrow NO_3^{-1}$ $H_2S \rightarrow SO_4^{-2}$

Although most hydrocarbon degradation is an aerobic process, anaerobic degradation occurs to a lesser degree in subsurface soils where oxygen levels

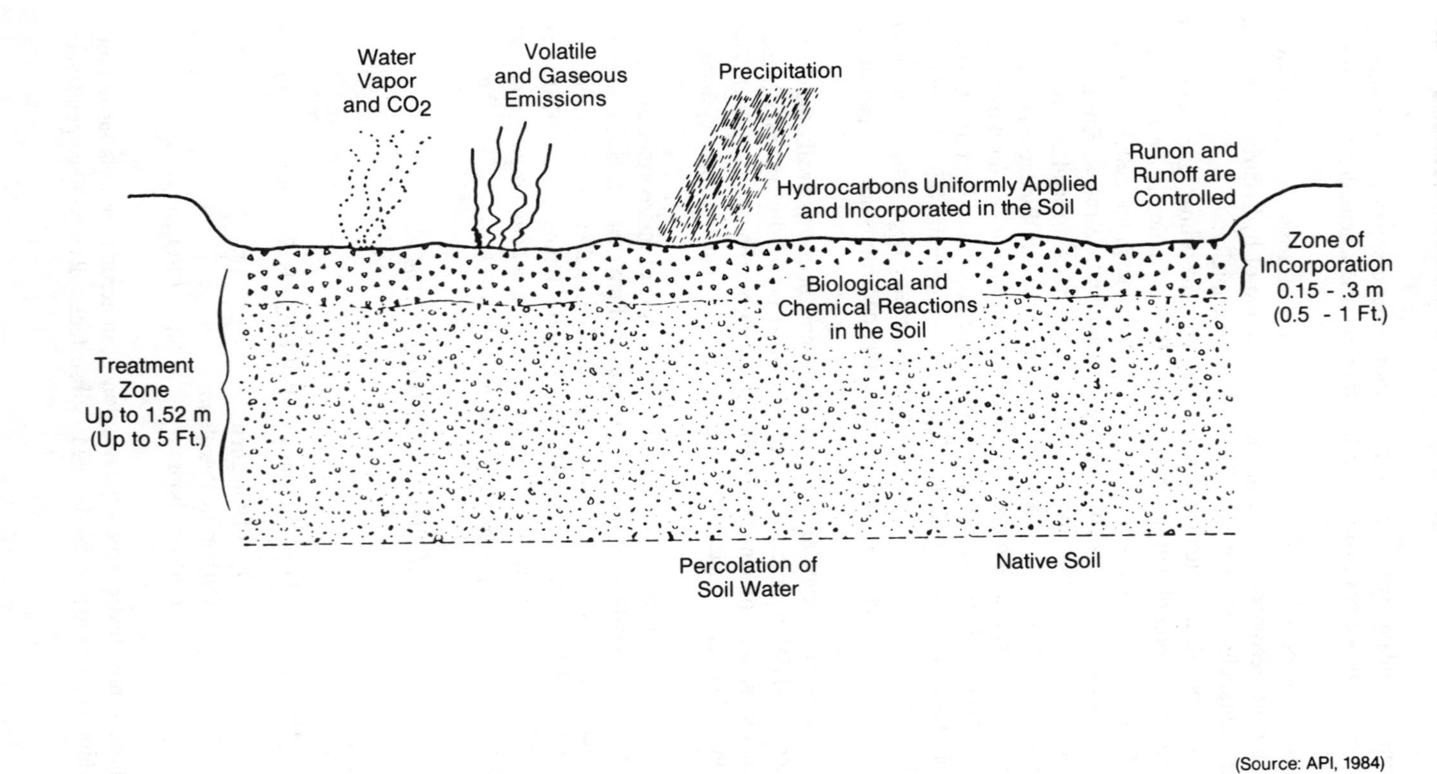

Figure 4.1 Mechanisms which occur during landfarming.

are decreased. Unlike aerobic degradation, however, anaerobic degradation of hydrocarbons is not complete. Anaerobic microbial metabolism of hydrocarbons promotes the accumulation of organic acids, alcohols, amines, and mercaptans as shown below:

ANAEROBIC BIODEGRADATION

Petroleum
Products
$(CHO)_nNS \dashrightarrow$ CO_2 + Microbial Cells + Organic Intermediates +

(20% of Original Carbon in Petroleum Products) (70% of Original Carbon in Petroleum Products)

$$CH_4 + H_2 + NH_4^{+1} + H_2S + H_2O + \text{Energy}$$

Aerobic petroleum biodegradation is discussed in detail in Section 3, In Situ Biodegradation. The metabolic processes are the same for in situ and non-in situ systems. The factors influencing biodegradation of hydrocarbons in soil are grouped into four categories: soil, environmental, chemical, and management (Table 4.1).

Table 4.1 Factors Influencing Biodegradation of Hydrocarbon Compounds

Soil Factors	Environmental Factors	Chemical Factors	Management Factors
Soil Microorganisms[1]		Soil pH	
Topography	Temperature	Nutrients	Waste Loading[2]
Moisture Content	Precipitation		Hydraulic Loading[3]
Soil Texture			Aeration[4]

[1]*Soil Microorganisms*—The soil must contain the microorganisms responsible for degradation. Brown, et al., 1983, determined that strains of *Pseudomonas, Arthrobacter, Flavobacterium,* and *Corynebacterium* are present in large numbers in oily wastes. Most soils normally contain these microbes.

[2]*Waste Loading*—Waste loading should be light, up to 5 percent by weight. Overloading the soil may result in toxic concentrations of hydrocarbons.

[3]*Hydraulic Loading*—Moisture content of the soil should be maintained at the level required to support microbial growth. This varies from site to site and also with the wastes applied. For very oily wastes, less water is required.

[4]*Aeration*—Aeration for the microbial processes is provided by tilling the soil. This is done after waste application and, if needed, at various intervals following the initial mixing.

Experience

Land Treatment Case Study 1—Kerosene. Dibble and Bartha (1979) demonstrated the effectiveness of landfarming in the cleanup of a major underground leak of kerosene. On 22 November 1975, near Crosswicks, New Jersey, an underground pipeline ruptured. For approximately 15 minutes over 1,890,000 liters (500,000 gal) of kerosene were sprayed, above ground, onto an area covering slightly more than 864.85 km^2 (3.5 acres). By 26 November, the standing surface oil was removed and approximately 198.9 m^3 (260 yd^3) of petroleum-laden soil were excavated and removed from the site of the leak. Landfarming techniques were utilized to treat the remaining kerosene (averaging 0.8 wt % oil) in the soils.

After the initial emergency responses, a groundwater collection system was installed and began operation by 3 December 1975. At that time, the first soil core samples were taken from two depths, 0 to 30 cm (0 to 11.8 in.) and 30 to 45 cm (11.8 to 17.7 in.) , at a point midway between the rupture point and the outer limits of the land treatment area. This sampling continued for the next 20 months.

The remedial program included a schedule of liming, fertilizing and frequent tilling of the sandy loam soil. In February 1976, the field was leveled and on 4 March, 10.41 kg/km^2 (5,660 lb/acre) of pulverized limestone were applied to the area. The next day the field was subsoiled to a depth of 46 cm (18.1 lb). The field was then tilled monthly through September 1976 to promote aeration. Fertilizer was applied on 8 March and again on 20 September at rates of .033 kg/km^2 (178 lb/acre) for nitrogen, 0.033 kg/km^2 (18 lb/acre) for phosphorus, and 0.028 kg/km^2 (15 lb/acre) for potassium.

The results of this procedure are summarized in Figure 4.2. Initially, oil concentration was higher in the upper 0 to 30 cm (0 to 11.8 in.) layer of soil and lower in the 30–45 cm (11.8 to 17.7 m) layer. However, after 6 months the relative concentrations were reversed due to the rapid decrease of oil in the upper layer. Enhanced biodegradation and volatilization were likely prime causes. Decreases in oil concentration also showed a strong correlation with temperature. The most rapid rate of biodegradation was attained when the temperature was 20°C (68°F) or higher in the months of July and August.

During the 21 months following the pipeline rupture, the concentration of petroleum hydrocarbons in the 0–30 cm (0–11.8 in.) upper soil layer decreased from an average of 0.80 percent to trace amounts. Kerosene was still found above trace levels in the lower soil layer, most likely due to the reduced aeration of that depth.

Land Treatment Case Study 2—Oily Sludges. Norris (1980) documented two successful landfarming operations at facilities owned by Imperial Oil

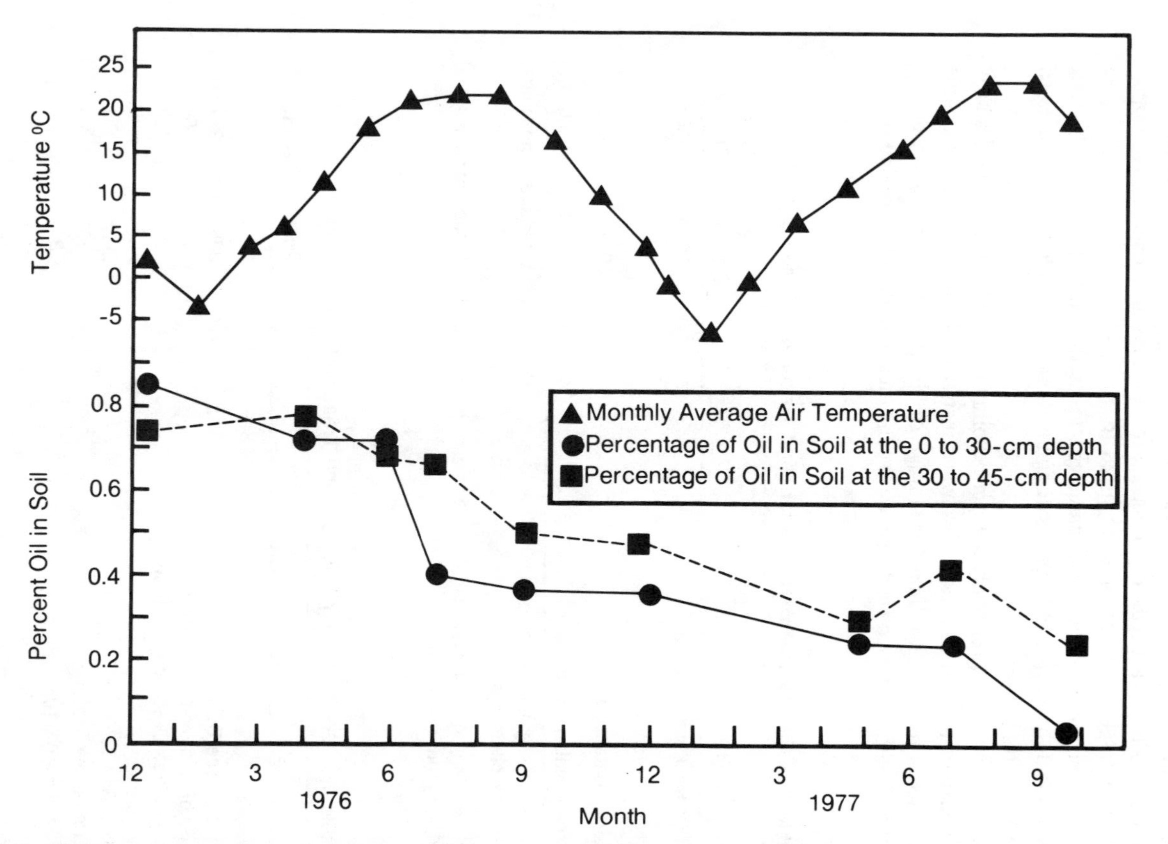

Figure 4.2 Air temperature and percentage of residual oil in soil following a pipeline break. (Source: Dibble and Bartha, 1979)

Refineries in Canada. In 1978, the Sarnia site began full-scale landfarming operations, while a test site was opened at the second site in Dartmouth.

Oily and biological (waste activated) sludges were treated at the two sites. Biological sludges are typically produced from wastewater treatment plant operations. Sludge characteristics are listed in Table 4.2.

Comparison of the two site characteristics is shown in Table 4.3. Neither site is ideal; however, appropriate design and operating practice has addressed the shortcomings.

At the Sarnia site, relatively low permeability soils prevented proper drainage of soil moisture so drainage tiles were installed. However, the same soils provided a natural clay liner for storage of sludge in shallow basins during winter months. Elevated roadways were constructed to provide access to storage

Table 4.2 Sludge Characteristics for Land Treatment—Case Study 2

	Oily Sludge[a]	Biological Sludge
Oil Content (wt%)	1–30	0.01
Solids Content (wt%)	1–20	1–10
Water Content (wt%)	50–98	90–99
Trace Metals (mg/kg solids)	1–10,000[b]	1–10,000[b]

[a]Primarily from fine oil removal (filter backwash, DAF froth, etc.), separator cleanings, and tank cleanings.

[b]Depending on the metal. Aluminum and zinc are among the highest; cadmium and molybdenum are among the lowest.

Table 4.3 Site Characteristics for Land Treatment—Case Study 2

	Sarnia	Dartmouth
Climate		
Frost-Free Days	235	237
Rainfall, May–October (cm)	45	60
Rain-Free Days, May–October	130	123
Topography	Flat	6-8%
Slope Soil Type	Clay	Silt
Permeability (cm/s)	10^{8}	10^{-5}–10^{-8}
CEC (meq/100 g)	30	4
pH	7–7.5	4.5–5
Depth to Bedrock (m)	15	30
Depth to Groundwater (m)	15	1–2

Source: Norris, 1980

and act as dikes to contain runoff on-site. Seven monitoring wells were installed to sample groundwater beneath the site.

At the Dartmouth site, trees, bushes, and rocks were cleared, and the land was graded to improve the slope. Six groundwater monitoring wells which surrounded the site and two within-site boundaries were installed. A concrete basin for winter storage was constructed. The pH of the soil was adjusted from 4.9 to greater than 7 by adding lime.

Fertilizer (sludge) was applied using different methods to each of the sites each spring. At Sarnia, sludge was distributed with a vacuum truck. After 6 to 10 truckloads (about 120–200 m^3/ha) the site was left to dry. A disk-harrow was then used to incorporate the sludge into the top 15–20 cm (5.9–7.9 in.) of soil. Aeration of the soil was performed biweekly or monthly by disk-harrow. A liquid manure spreader was used to spread sludge at Dartmouth and a disk harrow was then used to mix the soil with the sludge.

Several factors were monitored at both sites, but oil content and pH of the soil were the most important (Table 4.4). Oil content and pH were monitored no less than once a month and acted as guidelines for the landfarming operations. For example, if oil content was greater than 10 percent, application of the sludge ceased, or, if the pH was below 6.5, lime was added. Trace metals were monitored annually. At the beginning of each year, nitrogen and phosphorus levels were checked to assess whether enough fertilizer was present. Groundwater was monitored monthly. As expected by using this method, oil content was controlled and metals were retained in the mixing zone.

Land Treatment Case Study 3—Composting. Although composting of oily refinery sludges is practiced by a number of companies today, few publications exist which detail the composting of petroleum hydrocarbons. In December 1982, Texaco prepared a report titled "The Composting of Oily Sludges: Phase I and Phase II."

Table 4.4 Factors Monitored During Land Treatment—Case Study 2

Surface and Subsurface Soil	Groundwater
Oil Content	Oil
pH	Total Organic Carbon
Trace Metals	Phenol
Total Nitrogen and Phosphorus	Polyaromatic Hydrocarbons
	Trace Metals
	Nitrate, Ammonia, Phosphate
	Chloride, Sulfate, Conductivity
	Total Inorganic Carbon
	pH

Source: Norris, 1980

Texaco had conducted a pilot-scale study in 1977 which examined the composting of oily waste-activated sludges from various refineries. During Phase I, the oil content of compost in piles was reduced from 2 percent to 1 percent after 32 days.

In Phase II, respirometric and composting reactor studies were continued to determine the optimum temperature for degradation and to investigate the effect of composting on the leachability of metals from the sludge. The optimum temperature for biodegradation varied from 20°C (68°F) to 30°C (86°F). This temperature range correlates well with the optimum temperature range for landfarming of oily sludges. Leaching tests were performed for 14 metals. Some metals, such as barium, were unaffected; others, such as manganese and zinc, exhibited a decrease in leachability.

Implementation Feasibility for Land Treatment

Design Considerations

The design of a land treatment operation should consider characteristics of the material to be treated and characteristics of the proposed land treatment site.

Characteristics of Soil for Land Treatment. Design considerations are based upon the following soil characteristics:

- Soils applied to the landfarm should contain sufficient amounts of organic matter susceptible to biological degradation, be applied at a reasonable rate, and should not be prone to substantial leaching. Generally, soil laden with petroleum hydrocarbons has these characteristics.
- Applied soils should not contain high concentrations of constituents which could inhibit the growth of native soil microbes. Testing on a bench scale or field pilot scale can be conducted to assess compatibility with the native microbes.
- Applied soils should not contain leachable constituents which will adversely affect the quality of underlying groundwater. Nitrates are a concern. Petroleum hydrocarbons generally do not contain significant concentrations of nitrates. However they may contain other leachable constituents such as lighter fraction aromatic hydrocarbons (see Section 3, Environmental Feasibility of In Situ Volatilization).
- Applied soils should not contain materials which will significantly alter the soil's physical characteristics, including infiltration, percolation, and aeration potential. Petroleum products generally do not contain such materials, although excessive oil loading (application rate) to a landfarm can result in soil clogging and anaerobic conditions.

Considerations Related to Site Characteristics. The following characteristics should be considered when reviewing the suitability of a site for landfarming (American Petroleum Institute, 1980):

- Groundwater—The water table should be below the 1.5 meter (5 ft) treatment zone to provide sufficient protection of water quality.
- Soils—The soils should be moderately drained so that proper soil moisture (40 to 60 percent saturation) can be maintained in the treatment zone.
- Slope—A slope of 2 to 5 percent is typically considered acceptable.
- Cation Exchange Capacity (CEC)—The cation exchange capacity is an indicator of the ability of a soil to adsorb or retain cationic and/or metallic species. CEC is expressed in milliequivalents per 100 grams of soil (meq/100g) and can vary from 200 meq/100g for pure humus/organic matter to 4 meq/100g for hydrous oxide clays. A high CEC suggests that the soil can bind a large amount of cations and therefore provide more treatment capability.
- Location—The site should be located a sufficient distance from residential or public use areas to ensure that safety criteria are met and to provide an adequate buffer zone for any odors released.
- Utilities—No utilities should be situated above or below the ground.
- Climate—Areas which regularly receive heavy rainfall are not recommended for exposed landfarming due to the higher leaching potential and difficulty maintaining aerobic conditions. Extremely cold climates are also not recommended for landfarming due to difficulties associated with handling frozen ground and the reduction of microbial activity.

Process Considerations for Composting Operations. In addition to the considerations already discussed, composting operations require an evaluation of the following process parameters:

- Aeration—Convection currents combined with pile turning usually provide enough aeration in windrows to support acceptable biodegradation rates. In static piles or vessels, forced aeration is used. The flow rate must be adjusted to ensure proper oxygen levels.
- Moisture—The moisture content of the compost mixture should be between 40 to 60 percent to support microbial activity.
- Bulking agent—For petroleum-laden soils, the bulking agent dictates how dense the pile will be. A low density pile promotes aerobic conditions.
- Cover—For outdoor operations, covering the compost pile facilitates heat retention. The cover may consist of previously composted material, straw, wood chips, or other appropriate material.
- Ground—The compost pile should be isolated from the topsoil for outdoor operation. This prevents nondegraded contaminants from contacting and impacting soil or groundwater.

- Location—The location of the compost piles should be such that proper drainage away from the pile is provided during and after a rainstorm. Leachate from the pile may require treatment if contaminated. In general, composting systems do not produce leachate.

Equipment Requirements

Standard earth moving equipment is used for preparing the landfarm area and applying the material for treatment. The equipment should be selected according to the size of the operation, and may include such items as a dump truck, front-end loader, grader, or dozer.

Several methods are available for mixing or cultivating the soil. These alternatives are (Phung, et al., 1978):

- *Rototilling*. Rototillers provide a thorough mixing of the waste and soil. The standard rototiller is self-propelled with adjustable blades allowing a mixing depth of 46 cm (18 in.) and a mixing width of 2.4 meters (8 ft). Maintenance of a rototiller may be difficult and expensive due to the severe operating environment.
- *Disk Tilling*. The disk tiller, disk plow, and disk harrow all operate in the same manner to mix the waste with the soil. Large disks are lined in a row and rotate through the soil, promoting incorporation of the excavated soils into the treatment system soils. This method of mixing is not considered to be as thorough as a rototiller. However, because of sturdy construction, disks should perform well for a relatively long time.

In order to maintain proper soil moisture content, irrigation systems, with sprinklers, may be required if the site is located in a dry climate.

Composting requires equipment to mix the bulking agent and contaminated matrix, manipulate piles, aerate piles, and screen compost (if required). This equipment will vary depending upon the particular type of composting system used. Mixing equipment includes compost machines, front-end loaders, mixing boxes, pugmills, and agricultural rototillers. Aeration equipment requires stationary and moveable piping and blowers. Piles are generally manipulated with front-end loaders or windrow machines. Compost screening generally relies on various types of screens or a vibratory deck.

Treatment Needs

Treatment needs for compost operations are minimal. Proper aeration, moisture content, and nutrient content should be maintained throughout the process.

Soil samples which are collected from the treatment zone provide an indication of treatment effectiveness. Detection of a significant or increasing quantity of non-degraded organics suggests that the organics are not being properly degraded and that system modifications are needed. Reasons for this may be:

- The soil is overloaded with petroleum hydrocarbons because of an excessive application rate.
- Biodegradation ceases due to a low application rate, inappropriate moisture content, or lack of nutrients.
- Aerobic conditions are not maintained due to poor drainage, infrequent tilling, or overloading of the soil.
- Metals are migrating if the pH of the soil is too low.

Based on monitoring results, operational modifications and corrective measures can be prescribed to mitigate these problems.

Disposal Needs

Typically, the only significant sidestream from landfarming is surface runoff. Disposal of site runoff includes three options: evaporation, reapplication to the landfarm, and treatment at a wastewater facility.

Monitoring Requirements

Site and process parameters require monitoring to verify the effectiveness of land treatment operations and assure proper containment. The following parameters should be characterized and/or monitored:

- Background Information—Before landfarming operations commence, soil and groundwater samples should be analyzed. These samples will provide a reference for all future sampling.
- Unsaturated soil (vadose) zone—Monitoring of soils within and below the treatment zone will reveal if compounds are properly degraded or migrating through the vadose zone.
- Soil Pore Moisture—Sampling of moisture from the soil pores below the treatment zone provides an early indication of vertical migration of constituents within the treatment zone. This can be accomplished through the use of soil lysimeter monitoring devices.
- Groundwater—Groundwater monitoring will indicate whether compounds have leached to the groundwater. If so, corrective action may be required.
- Surface Runoff—Sampling of surface runoff may be necessary to assess water quality.
- Air—Perimeter air monitoring may be necessary to assess air quality de-

pending upon the type of hydrocarbon material (i.e., volatile fraction), climatic conditions, location of the landfarm site, and application rates.

Permitting Requirements

Permitting requirements will typically be dependent on the size and duration of the landfarming operations. Large operations of long duration would normally require a facility permit for land treatment from the appropriate state agency. Small capacity or short duration operations, such as a small single event spill, may require temporary approvals from state regulatory agencies to preclude time consuming permitting. Possible impacts to groundwater and surface water often determine the regulatory requirements related to the land treatment process.

Environmental Feasibility of Land Treatment

Exposure Pathways

During excavation, installation, operation, and monitoring of a land treatment system, primary pathways for direct human exposure include the following:

- Inhalation of vapors from the soil or groundwater.
- Skin contact with soil or groundwater.
- Particle inhalation and/or ingestion during system excavation, installation, and operation.

Plant uptake comprises a secondary environmental exposure pathway during installation, operation, and monitoring.

Environmental Effectiveness

The effectiveness of landfarming is highly dependent on site-specific conditions. As mentioned earlier, physical and chemical soil properties, site hydrogeology, ambient temperature, and a variety of other factors influence the effectiveness of landfarming. Years of experience with the landfarming of petroleum compounds confirm the following:

- Landfarming is an effective means of degrading hydrocarbon compounds. Lighter compounds, including constituents of gasoline, will be preferentially degraded and volatilized. Heavier compounds will degrade at a slower rate and become bound to soil particles.

- Continuous landfarming of petroleum-laden soils can result in accumulation of metals in the soil matrix.
- Ultimate degradation rates are site-dependent and cannot be predicted. As a result, waste application rates may be set by regulatory agencies without the aid of data from bench or pilot studies.

Economic Feasibility of Land Treatment*

Capital Costs

Capital costs for landfarming operations can be relatively inexpensive. If the proposed site and large equipment are already owned by the operator, then capital costs are significantly reduced and the initial expense is minimal. Items and costs are shown in Table 4.5. Rental of major equipment may be considered as another alternative.

Installation Costs

These costs are relatively low because the land is the treatment medium. Site preparation such as removal of trees, shrubs, rocks, and other debris may cost on the order of $0.55–$1.10/metric ton ($0.5–$1/ton) of treated material. The addition of lime and fertilizer may cost $0.55/metric ton ($0.5/ton) of treated material.

Table 4.5 Capital Costs for Land Treatment Operations**

Item	Cost
Land	$0.55–$1.1/metric ton treated material
Dump Truck	$80,000–$100,000
Tractor	$23,000
Rototiller	$17,000
Disc Harrow	$33,000
Sprinkler	$1,000

**Based on 1978 figures and multiplied by a cost index of 1.2.

*All costs are presented in 1986 U.S. dollars unless otherwise specified.

Operation and Maintenance Costs

Operations and maintenance costs are shown in Table 4.6. Soil and waste analysis are considered part of operations because results from these analyses often dictate further treatment operations.

Qualitative Ranking of Cost

Landfarming and composting are relatively inexpensive remedial technologies when compared with other options described in this report.

THERMAL TREATMENT

General Description

Thermal treatment is the process by which the affected soils are removed from the ground and exposed to excessive heat in one of the various types of incinerators available. During the incineration process, the unwanted compounds are volatilized and/or destroyed depending upon the intensity of the heat. Both low-level and high-level thermal incineration alternatives are discussed in this section. The facilities which are available for those operations, as well as their feasibilities, are discussed in detail in this subsection.

Experience has shown that the Thermal Treatment process can be effective for the removal of hydrocarbon compounds in soils. However, the costs associated with thermal treatment are still relatively high.

Table 4.6 Operations and Maintenance Costs for Land Treatment Operations*

Item	Cost (per metric ton treated material)
Cultivation and Site Operations	$1.65–$2.20
Material Transportation and Application	$9.35
Soil Analysis	$5.50

*Based on 1978 figures and multiplied by a cost index of 1.2.

Process Description

There are many types of solids-processing incinerators currently available for use. Waste materials are incinerated in rotating kilns, fixed kilns or hearths, rotating lime or cement kilns, asphalt plants, fluidized bed incinerators, and low temperature strippers.

Of the types of incinerators that will be discussed in detail, rotating kiln and fluidized bed are available as transportable units for on-site soils processing and as large-scale commercial facilities. The low temperature thermal stripper is available only as a transportable unit. Transportable units are delivered on flat-bed trailers in modules that are quickly assembled and installed. Usually, only hook-ups for utilities are required. Site-specific regulatory, technical, and economic factors will determine the choice of on- or off-site processing. The factors affecting the choice of process are discussed under Implementation Feasibility.

Rotating Kiln

Although differences exist, the principles applicable to rotating kilns are typical to other kiln techniques. Rotating kilns are widely used for the destruction of organic compounds because they can handle a variety of waste materials (physical consistency and chemical composition), provide agitation, and are available.

Figure 4.3 depicts a simple rotary kiln system. The operations represented are common to most rotary kiln systems. In a rotary kiln, solids are fed into a sloping kiln and are conveyed by gravity to the rear where the ashes are withdrawn. Off-gas is drawn through a scrubbing system prior to discharge. The basic operations are:

- Influent waste feed—Soils can be fed continuously via hopper or batch fed from fiber drums.
- Combustion air feed—Air can be fed through dedicated ducts or through the supplemental fuel burner.
- Fuel burner—Supplemental fuel can be burned as necessary to maintain proper combustion temperatures.
- Ash removal—Ash and slag are removed at the rear of the kiln or via a quench system.
- Scrubber—A liquid scrubber is used to cool the effluent gases, remove particulates, and neutralize any acids that may be produced.
- Effluent treatment—Scrubber water may need cooling, neutralization, clarification, or other treatment prior to discharge or recirculation.
- Afterburner—A large refractory-lined chamber is often used to ensure

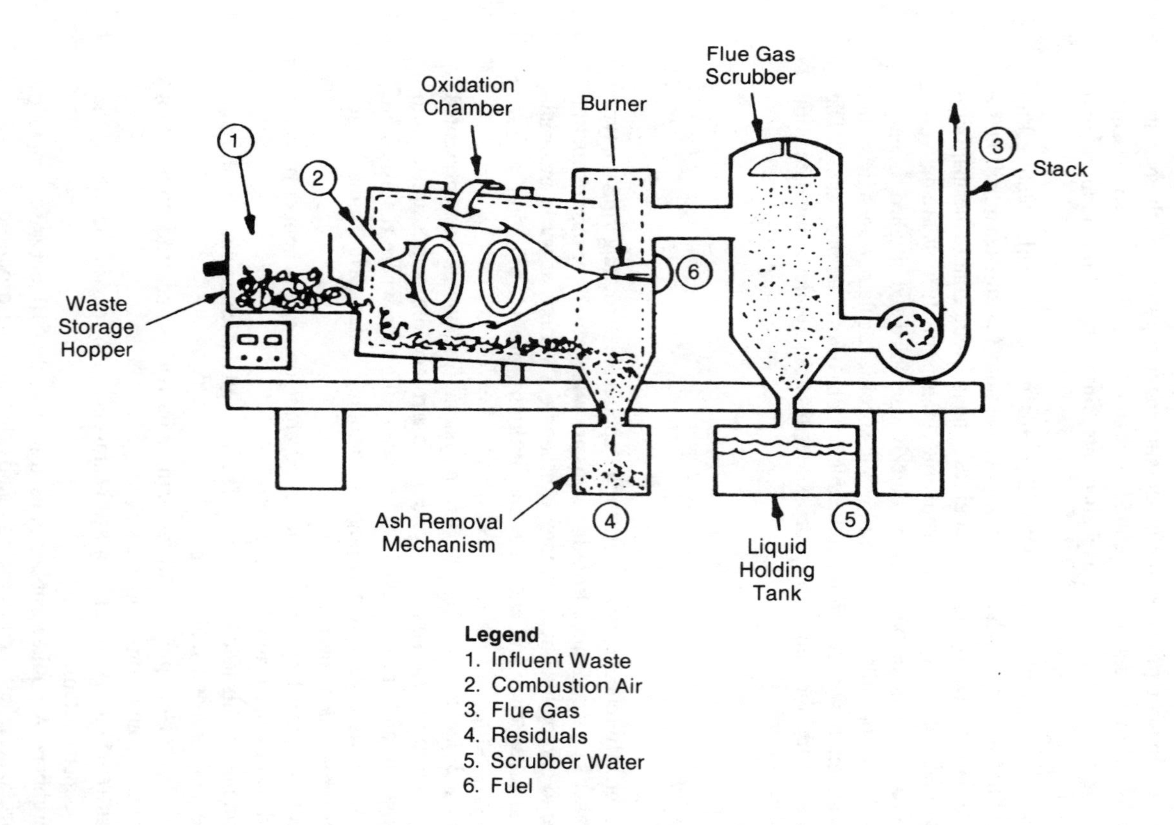

Source: Ghassami, Yu, and Quinlivan, 1981.

Figure 4.3 Rotary kiln incinerator schematic.

sufficient residence time at combustion temperatures for the complete combustion of any unburned organics in the combustion gases leaving the kiln.
- Stack—A stack is used to convey scrubbed gases to an elevation sufficient to ensure proper dispersion.

Fluidized Bed

A fluidized bed incinerator differs from a rotating kiln as it uses inert granular material (such as sand) for waste agitation and heat transfer. A typical fluidized bed incorporates a refractory-lined vessel in which the sand or other inerts are kept in turbulent motion. The waste solids are broken down by the sand and are removed as fine particulate while residual ash is removed at the base of the bed. Limestone can be added to neutralize acid gases formed as a product of combustion. Figure 4.4 illustrates a circulating bed combustor which utilizes the principles of fluidized bed technology (Vrable and Engler, 1986). In this design, finer inerts and higher bed velocities can be used by accommodating entrained inerts in a secondary chamber. The basic operations are:

- Waste feed—Soils are conveyed directly into the turbulent circulating bed.
- Forced draft (FD) fan—The FD fan is used to maintain air velocities which fluidize the bed.
- Cyclone—The inerts and/or off-gases are circulated through a cyclone which removes large particulates from the off-gas stream.
- Flue gas filter—Further reduction of particulates takes place downstream of the cyclone. A baghouse, electrostatic precipitator or equipment device can be used.
- Ash conveyor system—Ash is removed from the base of the fluidized bed.
- Induced draft (ID) fan—The ID fan creates a net negative pressure within the combustor which pulls the particulates from the fluidized bed.
- Supplemental fuel—For start-up and as needed, supplemental heat can be supplied by liquid or gas fuel feed and/or steam.

Low Temperature Thermal Stripper (LTTS)

The LTTS is a relatively new incineration technology which allows combustion of the hydrocarbon compounds without heating the soil matrix to combustion temperatures. In the LTTS, the soils are sufficiently heated in a twin-screw conveyor to volatize the organics. An indirect heat transfer fluid is used to heat the twin screw flights, but does not contact the soils. Volatiles are pulled off and incinerated in an afterburner. Figure 4.5 presents a schematic of the basic unit operations. The basic operations for the LTTS are described below.

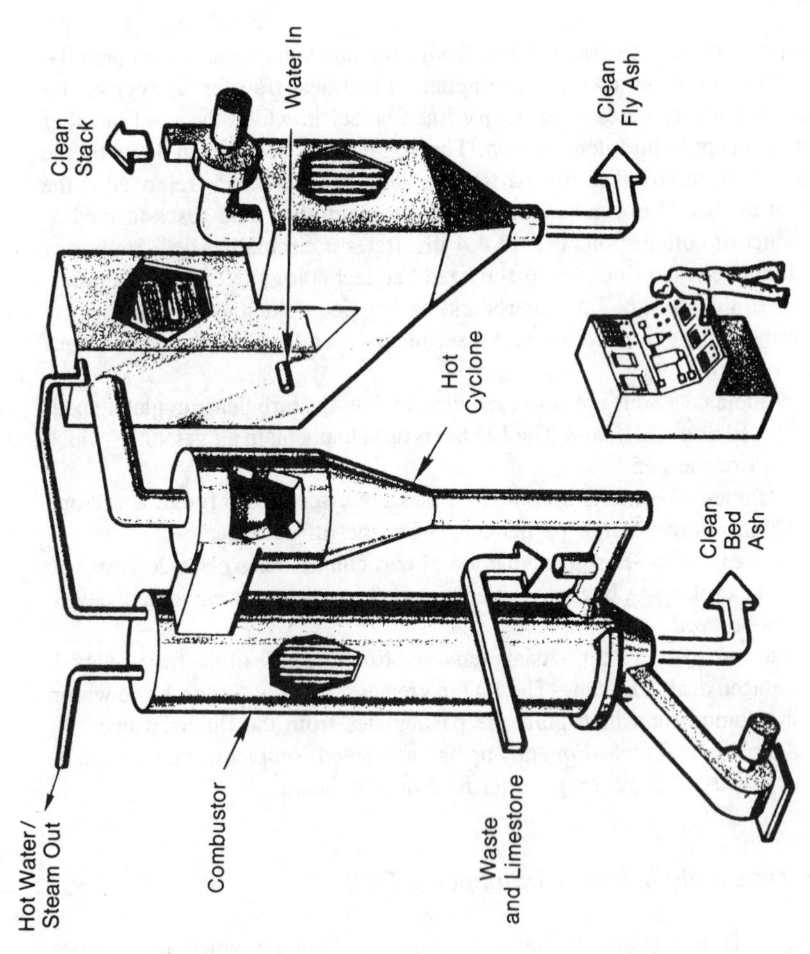

(Source: GA Corp, Undated)

Figure 4.4 An example of a circulating bed combustor.

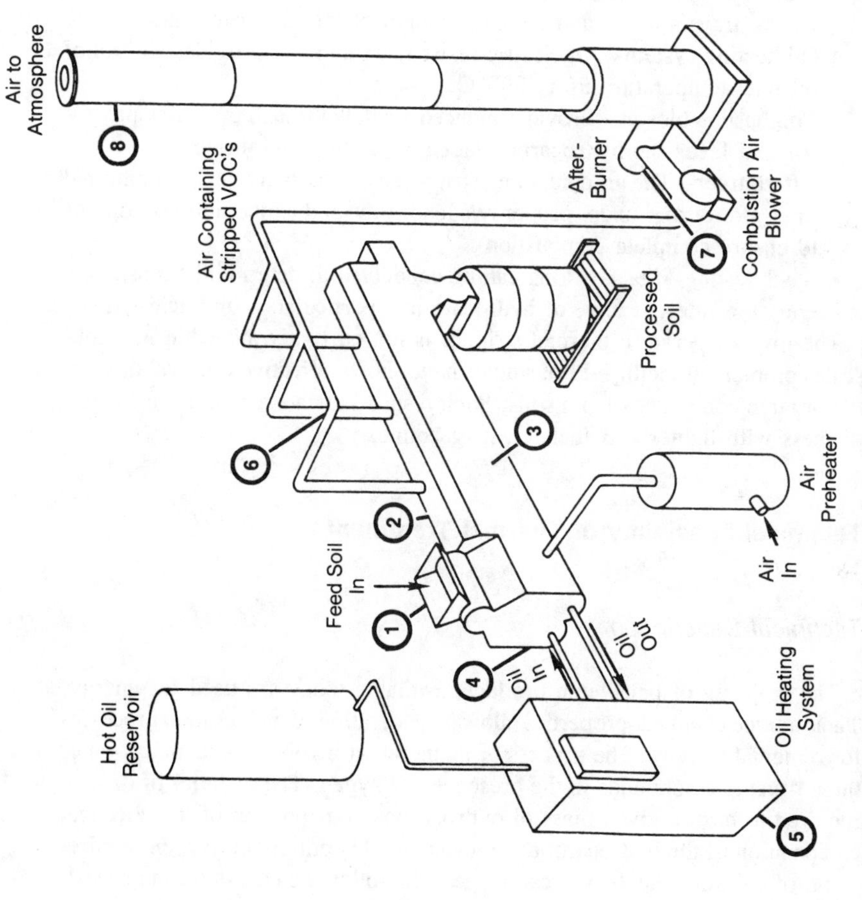

(Source: WESTON, 1986)

Figure 4.5 Schematic illustration of the low temperature thermal stripping system.

- Soil feed hopper—A rotary feed design provides positive feed control and ensures an air tight seal.
- Thermal processor—Air tight housing for twin-screw conveyor and heated trough.
- Screw system—Hot oil enters the first flight of each screw conveyor, travels the full length of each screw, then returns back through the center of each shaft.
- Trough jacket—Trough walls provide additional heat transfer surface. The hot oil travels down and back the length of the jacket trough.
- Oil heating system—An electric resistance oil heater provides control of oil inlet temperature up to 300°C.
- Combustion blower—Provides induced air flow through the screw processor and feeds the hydrocarbon laden air to the afterburner.
- Afterburner—The air containing stripped hydrocarbons can be combusted in an afterburner under proper residence time and temperature conditions to ensure complete combustion.
- Stack testing—Stack testing can be conducted to determine atmospheric emission rates, if any, of hydrocarbons, particulates, and acid gases.

The low temperature thermal stripper is not currently available in a full-scale commercial facility. Pilot studies have shown effective removal of volatile organic compounds from soils. Such results indicate a promise for effectiveness with lighter end fuels (e.g., gasoline).

Technical Feasibility of Thermal Treatment

Technical Description

The majority of petroleum products available today are used for energy. Their unique chemical properties allow consideration of incineration processes for remedial activity. The process of incineration involves the combustion of these organic compounds in the presence of oxygen. The quantity of oxygen required to burn a given mass of hydrocarbon is a function of the chemical composition of the hydrocarbon. To calculate this minimum oxygen requirement, often called the theoretical oxygen, the following equation can be used:

$$C_mH_n + ((4mn)/4)O_2 = mCO_2 + (n/2)H_2O$$

This equation depicts the complete combustion of a hydrocarbon C_mH_n in an oxygen atmosphere to carbon dioxide and water. The subscripts m and n are used to represent the number of carbon and hydrogen atoms present. During incineration, trace amounts of carbon monoxide, SO_x, NO_x, and gases and other by-products may be formed.

In typical combustion processes, theoretical air requirements do not meet actual minimum air requirements by an amount known as excess air. This extra air is determined by analysis of the products of combustion and is usually expressed as a percentage of the theoretical air. Ensuring sufficient air and air flow for proper combustion is a major consideration in the design of the incinerator.

In addition to oxygen requirements, the heat of combustion is a major consideration for the design and operation of thermal destruction processes. Minimum temperatures of operation mandate minimum energy requirements which must be met through the energy supplied by the hydrocarbon/soil feed and supplemental fuels.

Once the chemical composition of the hydrocarbon mixture has been determined and the mass (oxygen) balance and heat balance approximated, the incineration process and the operating conditions can be chosen. Destruction of hydrogen compounds is accomplished by exposing the hydrocarbon/soil matrix to an environment conducive to combustion. Parameters affecting this environment include:

- Air flow—Sufficient air must be in intimate contact with the combustible mass.
- Agitation—The mass must be agitated to continually expose new surfaces to combustion.
- Temperature—Minimum temperatures must be maintained to assure complete combustion.
- Feed rate—Excessive feed rate may cause material to pass through the incinerator without complete combustion. Feed rate may also be affected by Btu content, ash content, and metals content, as these elements influence the complete combustion of the hydrocarbon compounds in the soils.
- Residence time—Sufficient exposure time to elevated temperatures is required for complete combustion.

Knowledge of the chemical and physical parameters discussed above makes it possible to devise systems that can remove and incinerate hydrocarbons adhering to the soil matrix. These parameters and their interrelationships will be discussed in more detail under Design Considerations.

Experience

Incineration has been used for many years as a remedial technology, and the use of fluidized bed and low temperature thermal stripping has been recently increasing. Selected case studies are presented here. Although they are presented for specific chemical compounds, the techniques can be used for most volatile compounds.

Case Study 1—Rotary Kiln. Rotary kilns have been in use for many years and are proven technology. Coupled with an afterburner, rotary kilns can meet destruction efficiencies of 99.99 percent (also known as "four nines") for petroleum products. The use of transportable incinerators for the removal and destruction of hydrocarbons from soils is a more recent development. Pyro-Tech Systems of Tennessee, Resource Recovery of Louisiana, Ensco of Arkansas, and the EPA have developed mobile incinerators (Hankin, 1985). The EPA mobile incinerator is capable of handling up to 2.2 metric tons (2 tons) of soil per hour at 10 million Btu/hour, and reaches temperatures as high as 2,500°C (4,532°F) in the second combustion chamber.

The system can be transported on four semitrailers and consists of the rotary kiln as well as a secondary combustion chamber and air pollution control equipment. Replacement parts are transported with the incinerator on six additional trailers. In a 1982–83 test of this unit, the incinerator destroyed 99.9999 percent of the influent PCBs and 99.999999 percent of the carbon tetrachloride.

Based on these results, the use of this unit for more readily combusted petroleum products is promising.

Case Study 2—Circulating Bed Incinerator. Circulating bed incinerators or combustors (CBC) are employed internationally for a variety of users including waste incineration, steam, and electric generation. Commercial plants currently in operation are incinerating peat, wood, coal, oil and sewage sludge. A transportable unit is in service in San Diego, and another will go on line in the spring of 1987. The latter will be a 10 million Btu/hr model destined for initial use on a PCB-contaminated site in Alaska.

Rickman, Holder, and Young (1985) report the results of an extensive pilot program to demonstrate the effectiveness of circulating bed incinerators on a variety of influent streams. This pilot unit, a transportable model, comprises four separate combustion chambers and has undergone more than 7,500 hours of testing. The critical unit operation in this system is a 2 million Btu/hr combustor that can be configured in either a fluidized bed or a circulating bed mode. The destruction efficiency would be greater than 99.99 percent for petroleum products fed to this system as a saturated soil mixture.

Case Study 3—Low Temperature Thermal Stripping (LTTS). Noland, McDevitt, and Koltuniak (1986) demonstrated the use of a low temperature thermal stripper during a 22-day testing period in May 1985. The project was conducted by Roy F. Weston, Inc. (WESTON) for the U.S. Army Toxic and Hazardous Materials Agency (USATHAMA) to determine the feasibility of cleanup of a site contaminated with volatile and semi-volatile organic compounds.

The pilot-scale thermal processing equipment was transportable and was delivered to the site on a single flatbed truck. The entire system was installed and fully operational within 10 days after delivery to the site. During the 28 days of testing, more than 6,804 kg (15,000 lbs) of soil were processed. The VOC-enriched soil was excavated by backhoe from a site near the processing area. VOC levels in the soils were as high as 20,000 mgL̀ and were generally processed within hours of excavation.

The field demonstration program determined VOC removal efficiency under several varied operating conditions, including:

- Soil discharge temperatures (50°C to 150°C (122° to 302°F)).
- Soil residence times (30 minutes to 90 minutes).
- Air inlet temperatures (ambient to 90°C (194°F)).
- Heating oil conditions (100°C to 300°C (212° to 572°F)).

The air containing the volatilized organics was sent to the afterburner which provided a residence time of greater than two seconds at a temperature of 1,000°C (1,832°F) to ensure complete combustion of organics. The VOC-rich air served as the combustion air for the burner flame. Therefore, the organics were exposed to the high temperatures and turbulence within the flame vortex and complete combustion was ensured. Samples were collected for analysis from soil and off-gas to verify the destruction efficiency. In addition, temperatures, flows, and pressures were continuously monitored.

A comparison of the hydrocarbon concentrations measured in the feed soil to that measured in the processed soil and stack gas yield the following destruction and removal efficiencies:

- Greater than 99.99 percent removal of VOCs was evidenced in the soils.
- No VOCs were detected in the stack gas, indicating a destruction and removal efficiency (DRE) for the overall system greater than 99.99 percent.

Stack emissions were in compliance with all Federal and state regulations (including VOCs, HCl, CO and particulates). After regulatory approval, the processed soils were disposed of on-site as backfill.

The LTTS has not been tested for petroleum hydrocarbon removal. Experience with operating the unit indicates that the LTTS should be effective for the lighter fuels (i.e., gasoline) but less effective for fuel oils, diesel fuels, and similar heavy products.

Implementation Feasibility of Thermal Treatment

The following concerns were examined:

- Deciding between the use of an on-site transportable incinerator and an off-site commercial facility.
- Evaluating and establishing a transportable incinerator on-site.
- Utilizing the services of a commercial facility.

The following paragraphs describe design, equipment, treatment, permitting, and monitoring needs. Generally, the concepts and details addressed are applicable to each of the incineration techniques previously described.

Selection Criteria—On-Site vs. Off-Site

The decision to utilize on- or off-site incineration often hinges on more than cost considerations alone. The following factors, including cost, should be considered:

- Cost—The cost of the entire cleanup by both on-site and off-site approaches should be evaluated. A transportable unit can be very costly to site (see Section 4, Economic Feasibility, for detailed costs). However, once in place the operating costs in $ per metric ton can be modest compared to the cost per metric ton charged by commercial facilities. Generally, it is more cost effective to process smaller quantities of soils at an off-site facility.
- Utilities—On-site incinerators usually require electricity and water as a minimum. These utilities must be readily available in sufficient quantity to support the operation.
- Pollution control equipment—Both off-gases and scrubber water require treatment prior to discharge. The availability of a power plant boiler which can accept off-gases and/or the availability of a sewer system or an existing water treatment facility which can accept the scrubber water make on-site incineration more attractive.
- Permitting—An off-site commercial facility will already have all the permits required to incinerate wastes. Particular incinerators may have limitations on the type of materials that can be burned at that facility.

The decision to utilize on-site incineration technologies may not be an easy one. While the LTTS unit will work better for more volatile petroleum products, the fluidized bed and rotary kiln produce similar results. A variety of factors affect the decision, including soil agglomerate size (the fluidized bed may require size reduction), air pollution control (the rotary kiln requires an afterburner and pollution control unit), availability, ease of installation, utility requirements, permit requirements, and cost of a transportable unit. Often, the vendor of the transportable unit will operate the facility. In such cases, operational skills will be a factor.

Design Considerations

The design of any incineration system must provide for the removal and complete combustion of the influent petroleum products from the soil matrix. To accomplish this goal the following design considerations are necessary and are common to incineration systems:

- Air flow—To ensure complete combustion, sufficient air must be fed to the incinerator. This is often accomplished by feeding excess air to a supplemental fuel burner or establishing ports for air entry. A vacuum of a few inches of water is maintained to preclude possible "leakage" of combustion products outside the kiln.
- Agitation—Turbulence and mixing of influent solids can be provided by rotating conveyor screws, the rotation of the kiln, or by the fluidized bed. In all cases exposure of new surfaces for combustion is a critical parameter. Typical rotating kilns turn at a speed of approximately 30 rph.
- Temperature—The "rule of thumb" recognized by regulatory agencies is that complete combustion requires exposure of the volatilized hydrocarbon compounds to a temperature of at least 1,000°C (1,832°F) for a minimum of 2 seconds.
- Residence time—The two-second rule applies to hydrocarbon vaporized in the combustion chamber. An afterburner is often required to attain this two-second residence time after volatilization.
- Feed rate—Soil feed rate is a function of incineration system size, operating temperature, and system differential pressure (which determines residence time). A trial burn is a useful tool for empirically fixing feed rate. Soil feed to a rotating kiln is such that solids residence times are typically a few minutes.

The following restrictions or waste characteristics may require special design considerations:

- Soils may require repackaging in small 113.4 to 189 liter (30 to 50 gal) drums which can be batch fed to a kiln. Most commercial units are not equipped to handle bulk feed.
- Metals contamination (i.e., lead, mercury, etc.) of the soils may pose a problem for the incinerator effluent streams, ash, air, and scrubber liquor.
- Soils containing PCBs require specially permitted incinerators.
- Certain soil contaminants may also be restricted, i.e., chlorides, fluorides, sulfur, explosives, etc.

Equipment Requirements

Incinerators of differing design share common equipment requirements. The following list includes the major pieces of equipment for the incineration systems under discussion:

- The combustion chamber is generally refractory-lined to withstand the high combustion temperatures. The rotary kiln is a cylinder while the fluidized beds are generally rectangular. Ports for observation and liquid burners are built in.
- The solids feed system can be a conveyor belt which carries drums of soil to the incinerator opening or a hopper from which solids can be continuously fed.
- The air pollution control equipment is used to remove particulates and acid gases from the combustion air stream. A wet scrubber or dry cyclone are common. Limestone can be added in fluidized bed systems to preclude acid neutralization in the air pollution module. Because of the low pH waters generated in the quenching section of most air pollution control equipment, the use of Hastelloy C and reinforced fiberglass is not uncommon.
- The induction draft fan, or ID fan, is located on the tail end of the incineration system just prior to the stack. This fan draws combustion air through the air pollution equipment.
- The forced draft fan, or FD fan, is used to supply combustion air to the burners.
- Combustion burners are used with waste or supplemental fuels to heat the incinerator to operating temperatures and maintain those temperatures. In the low temperature thermal stripper the combustion burner is used to destroy hydrocarbons stripped from the soils in the thermal processing unit.
- A variety of monitoring equipment is used to determine and track incinerator parameters. Some of these parameters are stack gas emissions, incinerator temperature, combustion chamber pressure and scrubber water flow rate. More detail on these parameters can be found under Monitoring Requirements.

Specialized equipment may be necessary to excavate the soils. If transport to an off-site incinerator is contemplated a licensed waste hauler will be required. Depending on soil quantities, repackaging facilities and/or labor may also be necessary, prior to shipment, for ease of handling at the incinerator site.

Treatment Needs

Many Federal and state regulatory agencies have placed restrictions on the operation of waste incinerators. The definition of petroleum saturated soils as a hazardous or nonhazardous waste will, in part, determine the requirements for operating an on- or off-site incinerator. There is currently no Federal definition. Petroleum products are not listed as a hazardous waste under the Resource Conservation and Recovery Act (RCRA); therefore, some states do not consider petroleum-saturated soils a hazardous waste. The soils are also excluded from the Comprehensive Environmental Response Compensation and Liability Act (CERCLA). However, certain chemical constituents of petroleum products, such as benzene, toluene and xylene are considered hazardous under RCRA, and, therefore, some states consider petroleum-saturated soils hazardous (Kostecki et al., 1985). Judgment is on a case-by-case basis.

Disposal Needs

The by-products of incineration and their disposal needs are listed below:

- Incinerator ash and dry (cyclone) solids—Solids generated by the incineration of hydrocarbon compounds may be considered by regulatory agencies to be a hazardous waste. If so, landfilling in a secure landfill will be necessary. If not considered a hazardous waste, the on-site disposal-backfill may be an available and preferred option.
- Wet scrubber sludge—Sludge generated from the air pollution control equipment of the incinerator may be considered a hazardous waste particularly if contaminated with heavy metals. Laboratory analysis may be required to determine whether the material can be delisted as a hazardous waste.
- Scrubber water effluent—Commercial facilities have treatment plants to pre-treat the scrubber water prior to discharge. Transportable units must address predischarge treatment.

Monitoring Requirements

Table 4.7 details the monitoring and inspection requirements stipulated by the Code of Federal Regulations (CFR 265.347) when incinerating a hazardous waste. Most commercial and transportable incinerators are designed and permitted for hazardous waste use. The incineration of nonhazardous waste may allow for less stringent monitoring.

Permitting Requirements

Although not listed as an option in this section due to its impracticality for most site-specific purposes, stationary incinerators require many permits. Ob-

Table 4.7 Monitoring and Inspection Requirements

Monitoring Parameter	Frequency of Monitoring
Combustion Temperature	Continuous
Waste Feed Rate	Continuous
Combustion Gas Velocity	Continuous
Carbon Monoxide (between the combustion zone and release)	Continuous
Ash Composition	Upon request of EPA Regional Administrator
Exhaust Emissions	Upon Request of EPA Regional Administrator
Waste Feed Chemical Analysis	Determined by EPA Regional Administrator
Inspection Parameter	**Frequency of Inspection**
Incinerator and Associated Equipment	Daily
Emergency Waste Feed Cutoff System and Alarms	Weekly (or Monthly with Regional Administrator consent)

taining the permits for this type of facility has proven to be a lengthy and often costly process. Applications for permits to construct must be made to Federal, state, and often local agencies. Once built, a plan for a test burn must be submitted and approved. The test burn must show incinerator compliance with operations and emission requirements. A temporary operating authorization (TOA) may then allow the incinerator to begin operation. The permitting process for a temporary transportable incinerator may be less severe but is dependent on the state and locality. The transportable CBC unit in San Diego was required to meet local air requirements, has obtained an EPA permit for incineration of PCBs and is in the process of obtaining a RCRA Research, Development and Demonstration (RD&D) permit. Permits for water use and discharge will most likely be required.

Environmental Feasibility of Thermal Treatment

Exposure Pathways

Possible primary pathways for direct human exposure during excavation, installation, operation, and monitoring of a thermal treatment unit include the following:

- Inhalation of vapors from the soil or groundwater during excavation and installation, and/or inhalation of emissions from the treatment unit during operations and monitoring.
- Particle inhalation and ingestion during excavation system installation, and system operations.
- Skin contact with soils and groundwater during excavation and installation, and contact with the treated stream during operations. Care must be taken during material handling of process feed, ash, scrubber sludge, and neutralizing chemicals (i.e., caustic).

Possible primary pathways for environmental exposure during excavation are groundwater and surface water. The degree to which groundwater and surface water are affected often determines the regulatory cleanup levels.

No secondary exposure pathways have been identified for this process (see Table 2.5).

Environmental Effectiveness

The effectiveness of high temperature incineration for destruction of petroleum products is well documented. Destruction and removal efficiencies of 99 percent can be expected. The effectiveness of low temperature thermal stripping is dependent on the volatility of the particular hydrocarbon compounds present. It has been determined that even semi-volatiles, such as xylene, can be successfully thermal stripped and incinerated utilizing the LTTS system. Low temperature thermal stripping effectiveness is more sensitive to the type of hydrocarbon.

Economic Feasibility of Thermal Treatment*

Capital Costs

Capital costs for the typical incineration system can be high but vary widely

*All costs are presented in 1986 U.S. dollars unless otherwise specified.

depending upon design chosen and operating feed rate. The capital cost for a permanent low temperature thermal stripper can be as low as $1 to $2 million for a system that can process 550 metric tons (500 tons) per week. An additional $200 to $500 thousand would be required for an exhaust gas treatment system.

The capital costs required to establish a full-scale permanent incinerator will vary with design, capacity, and effluent treatment requirements, but will typically run in excess of $1 to $3 million. Once established, however, the costs for operation and maintenance are much lower than for a leased transportable unit, which will most likely be operated by the vendor. In general, smaller quantities of soils would be more cost effectively processed by a commercial facility, and very large spills may make the leasing and establishment of an on-site unit more attractive.

Installation Costs

The typical transportable rotary kiln with afterburner and associated pollution control equipment has set-up costs of $100–$500,000 depending on the need to clear land, build temporary pads, establish handling areas, and tie-in utilities. A transportable low temperature thermal stripper would have similar set-up costs. Another option is the GA Technologies mobile fluidized bed incinerator. This unit can be set up on-site for approximately $300,000 and will cost from $110 to $330/metric ton ($100 to $300/ton) of soil to operate (including excavation, pretreatment size reduction, if necessary, labor, and maintenance). Landfilling of ash, if required, is not included.

Operation and Maintenance Costs

Operating costs for a rotary kiln, fluidized bed, or thermal stripper can be high. The operating costs for a full-scale permanent rotary kiln have been estimated to be more than $250 thousand per year including auxiliary fuel, utilities and refractory replacement (EPA 1985). This value however includes $1.2 million for recovered steam. Without steam recovery operating costs rise to over $1.4 million per year. Maintenance costs are estimated to be approximately 10 percent of total equipment costs.

Operating and maintenance costs for a transportable fluidized bed incinerator vary greatly with design and mode of operation. However, an estimate of the costs is approximately $110/metric ton ($100/ton) for a unit that will handle 3.6 to 4.5 metric tons/hr (4 to 5 tons) of soil.

The O&M costs for thermal stripping also vary with size and design (i.e., need for exhaust gas treatment). However in this case, the fuel and electrical requirements greatly increase the operating costs. For thermal stripping, O&M costs can approach total equipment costs.

The cost for incineration at a commercially-operated facility could be as high as $85 per fiber-packed 75.6 liter (20 gal) drum or the equivalent of over $1,100/metric ton ($1,000/ton).

Qualitative Ranking of Cost

Incineration options rank among the most costly remedial activities. The use of an existing permitted facility is very expensive. Utilizing a mobile incinerator, which may be rented, may lower the cost.

ASPHALT INCORPORATION

General Description

Asphalt incorporation is a recently developed remedial technique that involves the incorporation of petroleum-laden soils into hot asphalt mixes as a partial substitute for stone aggregate. This mixture can then be utilized for paving.

During the incorporation process, the mixture, including the affected soils, is heated. This causes volatilization of the more-volatile hydrocarbon compounds at various temperatures. The remainder of the compounds become incorporated into the asphalt matrix during cooling, thereby limiting compound migration. This process, as well as the feasibility of this process, is described in detail within this subsection.

Asphalt incorporation is a relatively new technology. Little is known about the effects of various types of hydrocarbons on the admixtures of asphalt. The work that has been done with this process has been conducted with virgin, not waste, hydrocarbons and with very small amounts of soils. To date, only one regulatory agency has allowed Asphalt Incorporation of affected soils.

Process Description

Figure 4.6 details a typical hot mix asphalt batching operation (Asphalt Institute, 1983). Unheated aggregates stored in cold bins (1) are proportioned

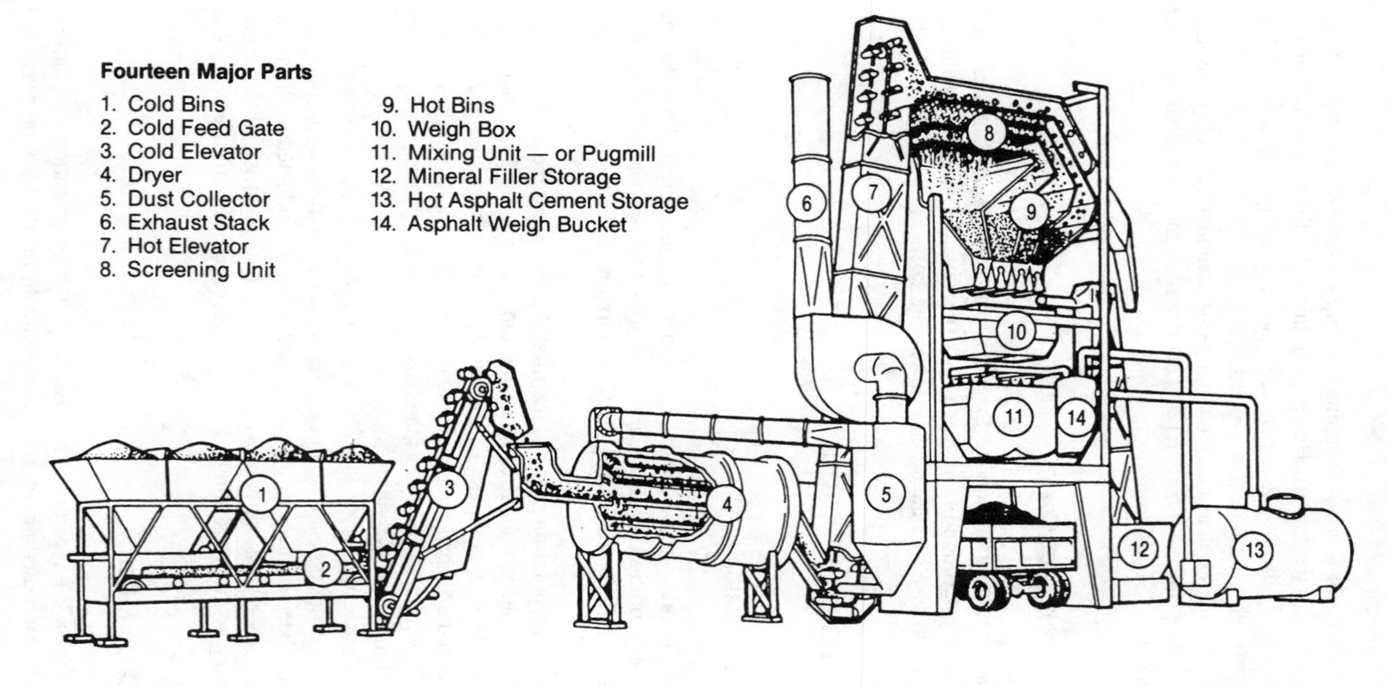

(From: Asphalt Institute, 1983)

Figure 4.6 Typical asphalt batching plant

by cold feed gates (2) onto a conveyer system (3). The conveyer system delivers the material to a dryer (4) where it is heated and dried. The dust collector (5) and the exhaust system (6) remove exhaust gases from the dryer. In a more modern plant than the one pictured, aggregates are then delivered by a hot elevator (7) to a screening unit (8) which separates the material into different fractions by particle size and deposits them into hot storage bins (9). The aggregates are measured into the weigh box (10) and are then dumped into the mixing chamber (11), along with any mineral filler that may be needed (12). Hot asphalt is then pumped from storage (13), weighed (14), and delivered to the mixing chamber where it is completely mixed with the aggregates. The final product is then dumped in a truck or conveyed to heated storage.

When soils containing petroleum products are utilized in a hot mix asphalt batching plant, the soils are simply added to the aggregate feed stream as a small percentage of the total feed. In practice, the soil feed should be limited to less than 5 percent of the total aggregate feed at any one time. This restriction is required in order to maintain final product quality and to minimize air emissions caused by volatilization of the compounds.

Technical Feasibility of Asphalt Incorporation

Technical Description

The primary mechanisms of remediation for asphalt incorporation are thermal destruction by incineration and volatilization. During the asphalt batching process, aggregate is passed through a dryer where average temperatures range from 260°C to 427°C (500°F to 800°F). Retention times approach five minutes. Aggregate leaves the dryer at approximately 149°C (300°F). When petroleum bearing soils are utilized in an asphalt batching plant, volatile organic compounds are volatilized within the asphalt dryer.

Large quantities of the lighter petroleum hydrocarbons contained in gasolines, kerosenes and fuel oils are incompatible with asphalt. These compounds can act as solvents to soften the final asphalt product. Heavier fractions are chemically similar in nature to asphalt and may not be as damaging to the asphalt mix. Solidification or encapsulation within the asphalt/aggregate matrix is a secondary mechanism of soil remediation which serves to contain heavier hydrocarbon fractions that might not be removed during the drying process.

Experience

The incorporation of soils containing petroleum products into asphalt mixes is not widely practiced. A review of the current literature on asphalt mixes and additives found no mention of this process. Two asphalt industry representatives who were interviewed reported that it is not considered to be common practice (Joubert, Pagan, 1985). However, a national survey of state agencies has found that five states (Massachusetts, Minnesota, Rhode Island, New Hampshire, and Vermont) have had experience with the disposal of petroleum-laden soils by this method (Kostecki et al., 1985).

In 1984, the Massachusetts Department of Environmental Quality Engineering approved an application from an asphalt batching plant near Worcester, Massachusetts, which allows them to use petroleum-bearing soils as part of their mineral aggregate feed. The soil is incorporated at less than 5 percent by weight of the total feed. Even at this small percentage, this plant has the capacity to use 7,273 metric tons (8,000 tons) of soil annually.

The approval letter regarding this application specifies strict management guidelines which must be followed by the plant. The plant was required to construct a sloped concrete pad that was capable of holding 765 cubic meters (1,000 yd^3) of soil in storage. The pad was surrounded by a dike, lined with a coal tar emulsion sealer, to prevent possible leaching of petroleum products into surrounding soils. The pad and storage pile must be covered when the plant is not in operation. Any runoff from the pad is collected and periodically removed to a hazardous waste treatment facility.

In addition, the plant may only accept soils that contain virgin fuel products. Used and waste oils could contain unknown constituents that would pose a health threat to workers and occupants of the surrounding area. Only waste material generated in Massachusetts can be accepted at the site. During the winter months, when the plant is closed, no hydrocarbon-bearing material can be kept in storage at the plant site. The plant must also comply with a strict manifest and record-keeping system, even though the soil is not considered to be a hazardous waste under the terms of the application approval.

This method of disposal for petroleum-laden soils is now considered to be the most favored option in the State of Massachusetts. State officials plan to encourage at least one operator in each of the state's four regions to retrofit their plants in a manner that is similar to that described above.

Implementation Feasibility of Asphalt Incorporation

The decision to implement this technology requires an examination of the following considerations:

- The soils must be cost-effectively excavated. If soils cannot be readily excavated and transported, this option, as well as all off-site options, will be difficult.
- Soil should be incorporated into the mix at less than 5 percent of the total aggregate feed. This is to ensure that the amount of fine material in the aggregate feed will be less than 10 percent of the total aggregate feed.
- The soil must be free of large rocks, wood, and debris.
- At this time, the technology is accepted only in a few states and, therefore, is not widely available.
- Asphalting operations are seasonal and do not operate in cold weather.

Design Considerations

This subsection presents several of the technical uncertainties associated with asphalt incorporation. Because the technology is relatively new, these types of uncertainties will be resolved as the technology gains further acceptability.

In theory, the high temperatures encountered by petroleum-bearing soils in the aggregate dryer should completely volatilize all volatile organic compounds (VOCs) that might be present; however, no experiments have been performed to verify this. Proponents of this disposal option point out that secondary stabilization is provided by the asphalt mix because the soil is ultimately locked into the asphalt/aggregate matrix. Asphalt industry representatives and geotechnical engineers have pointed out that incomplete destruction of the fuel hydrocarbons, as well as characteristics of the soil itself, could have deleterious consequences for the final asphalt product that must be considered in the design and operation of such a system (Bemben, Joubert, and Pagan, 1985).

Any residual hydrocarbons that are not burned off in the dryer could soften the mix, as when asphalt is ''cut'' with fuel oil, kerosene, or gasoline, to create slow, medium, or rapid curing asphalt. Soil requiring remediation that is stored in bulk may contain all three of these substances so it is possible that curing times for different portions of the same mix may be unpredictable.

The strength and durability of asphalt paving mixes is dependent on the size, type, and amounts of aggregates used. Normally, the amount of fine material which passes the No. 200 sieve is limited to 2 to 10 percent of the mixture. As the size of the particle decreases it tends to absorb a larger percentage of the hot asphalt due to increasing surface area. Too many fines could lead to coating problems or a mixture that is too dry. In an extreme case, this imbalance could lead to cracking or unstable pavement. Further, clays and soil organic matter should be absent from asphalt cements. Petroleum-bearing soils from mixed sources can vary considerably in quality and could easily contain soils which could damage the final asphalt mixture.

Unfractured stones within the mix could make the product less stable. In the Massachusetts example cited above, the final product cannot meet specifications set by the Massachusetts Department of Public Works for use on state highways due to the amount of unfractured material in the mix. The mix can be used for all other paving purposes. Screening the soil may successfully remove unfractured stone size particles.

Equipment Requirements

Standard excavation, earth-moving, and transporting equipment will be required to utilize this option (see Chemical Extraction).

When petroleum-laden soils are utilized in an asphalt batching plant, additional equipment is needed. Specifically, an additional cold storage bin or hopper is needed, as well as a system for conveying the soil to the rotary dryer. The conveyer system must provide some means of metering the amount of soil that is sent to the dryer. Facilities for storage of the material must also be constructed. This equipment is simple to design. It may, however, require a large capital expense, depending on the size and configuration of the asphalt plant that is being retrofitted.

Treatment Needs

The treatment needs for this technology are not completely known at this time. Depending upon the on-site soil characteristics and asphalt plant aggregate specifications, pre-screening of the soil material may be required to remove rocks, sticks, clays or fine-grained material. State-specific regulations may also require an asphalt plant to upgrade its emission control system to handle volatiles released in the dryer.

Disposal Needs

There are two main disposal needs that must be addressed when one is considering the reuse of soils that contain petroleum hydrocarbons via the asphalt batching process. The first involves uncertainties concerning the actual extent to which hydrocarbon compounds are destroyed via incineration or are volatilized to the atmosphere.

The second disposal need was discussed in detail under Design Considerations and concerns the effects that petroleum-laden soils and ungraded soils in general could have on the integrity of the final asphalt mix. Both needs impose practical limits on the quantity of soil that can be included in the aggregate feed.

Monitoring Requirements

A hot mix asphalt batching operation is a complex process which requires constant monitoring. A plant that incorporates petroleum-bearing soils into its mix may require additional monitoring. Plant operators must make sure that the soil does not adversely affect the product mix or the plant air emissions quality. In addition, regulators may require that soils stored on site be sampled periodically to ensure compliance with the virgin product rule.

Permitting Requirements

Permitting requirements for this disposal option will vary on a state-by-state basis. The asphalt industry is already highly regulated by transportation and air quality regulators. Asphalt plant operators must comply with the permitting requirements of these agencies and may face additional permitting requirements if their plant is considered to be a soil treatment facility. The extensive regulatory requirements faced by asphalt plant operators may have considerable bearing on their willingness to accept petroleum-laden soils for processing.

Environmental Feasibility of Asphalt Incorporation

Exposure Pathways

Asphalt incorporation of soils containing petroleum hydrocarbons offers the advantage of soil remediation via volatilization and incineration of hydrocarbons contained in those soils. As was mentioned in the case study, there is concern that incomplete combustion of volatile organic compounds could compromise air quality in the vicinity of the plant. The extensive air pollution control equipment normally required for hot mix asphalt batching plants and the restrictions on the allowable percentage of soil contained in the feed helps alleviate this concern.

During excavation, operation, and monitoring of an asphalt incorporation system, possible primary exposure pathways for direct human contact include the following:

- Inhalation of vapors from the soil or groundwater during excavation and

inhalation of emissions from the treatment unit during operations and monitoring.
- Particle inhalation and/or ingestion during excavation and system operation.
- Contact of skin with soil or groundwater during excavation and system operation.

Possible primary pathways for environmental exposure during excavation and operation of an asphalt incorporation system are groundwater and surface water. Impacts to the groundwater and surface water often determine the regulatory cleanup levels if these receptors are affected at the site. Water ingestion could result if the soil is stockpiled in a manner that allows water to leach through the material. A concrete storage pad and provisions for covering the storage pile during the hours when the plant is closed can help prevent the generation of runoff from the storage pile which might contain hydrocarbon constituents.

No secondary exposure pathways have been identified for this process (see Table 2.5).

Environmental Effectiveness

The time required to dispose of hydrocarbon-laden material through the asphalt batching process is limited only by the size of the batching plant. Material may be excavated and stored until it can be used.

As stated previously, acceptable material should be limited to that containing only virgin hydrocarbon product. This is a quality control and safety measure, required because the material often arrives at the site in small quantities from various sources.

At present, no measure of hydrocarbon removal efficiency is utilized. This could easily be accomplished by testing the aggregate for hydrocarbon content after it exits the dryer. Presently, it is assumed that the high temperatures encountered in the dryer and the secondary encapsulation of the soil within the asphalt mix provide adequate remediation measures.

Economic Feasibility of Asphalt Incorporation*

Capital Costs

The cost of retrofitting an asphalt batching plant to accept soils containing petroleum products is borne by the plant operator. This cost is likely to be

*All costs are presented in 1986 U.S. dollars unless otherwise specified.

on the order of tens of thousands of dollars and is offset by the fees that are charged to the soil generators.

Capital costs for excavation and transportation are minimal as equipment and labor for these tasks are typically contracted.

Installation Costs

Installation costs include the cost of installing the retrofitted equipment. This cost is also borne by the operator who usually maintains a crew that is capable of performing customized modifications to the existing plant structure. Other workers may need to be hired to design and build structures such as the concrete storage pad and runoff collection system.

Operation and Maintenance Costs

Maintenance of the equipment is an additional expense; however, the operations involved may be the same as those required for the other plant equipment. Additional recordkeeping and reporting may be required. These are a function of state and local regulations and will differ from site to site.

Fees charged to generators of petroleum-laden soils are expected to be lower compared to other forms of off-site disposal. The average charge is currently around $55 to $82.50 per metric ton ($50 to $75 per ton) of material (1985 costs). Excavation costs (see Environmental Feasibility of Excavation) are the same as for other off-site disposal methods.

Qualitative Ranking of Cost

The costs involved in this option are a function of the ease of excavation. Should excavation prove to be straightforward, the costs will be comparable to or less than other off-site technologies.

SOLIDIFICATION/STABILIZATION

General Description

Solidification/stabilization processes have received little attention as potential remedial actions for soils that contain petroleum products. However, there may be specific instances where the use of these techniques is applicable. Solidification/stabilization is the process by which an additive is incorporated into the excavated soils to encapsulate the compounds of concern. Through this process, the contaminants are bound to the soil/additive matrix to prevent their migration. This provides safe disposal or reuse of the treated soils. This process, as well as its feasibility, is discussed in detail in this subsection.

Solidification/stabilization, which encapsulates the hydrocarbons within the soils, has not been widely utilized because ultimate destruction of the compounds does not occur.

Process Description

Solidification/stabilization processes can be performed either on- or off-site. These processes involve only the mixing of the material to be disposed with the various stabilizers and additives. Figure 4.7 depicts a generalized process for the manufacture of pozzolanic material utilizing fly ash (Transportation Research Board, 1976). Figure 4.8 is a flowchart of the Envirosafe pozzolanic waste solidification process which uses fly ash (Smith and Zenobia, 1982). Material solidified by this process is transferred to an off-site location to be processed.

Solidification/stabilization processes have more commonly been used to stabilize oily wastes and sludges contained in surface impoundments. Surface impoundments can be stabilized in two ways. The first is in situ. The stabilizing agent is added directly to the impoundment and thoroughly mixed. The impoundment is generally treated in small sections. As one section solidifies it is used as a base which allows the equipment to reach further out into the impoundment. Figure 4.9 details the stepwise treatment of a typical impoundment (Musser and Smith, 1984).

The second method involves excavation of the sludges contained in the impoundment. This is sometimes called the area mixing method. The steps involved in the solidification process are as follows (Morgan et al., 1984):

- Bulldozers level piles of kiln dust into 15.24 to 30.48 cm (6 to 12-inch) deep layers.
- A backhoe lifts the sludge from the impoundment and places it on top of the kiln dust.
- A bulldozer then mixes the two materials together, and a pulverizing mixer is driven over the mixture until homogeneity is achieved.
- The mixture is allowed to air dry for about 24 hours and is then compacted and field tested.

The layers can be stacked to build an in-place landfill, or the semisolidified sludge can be trucked to another landfill location.

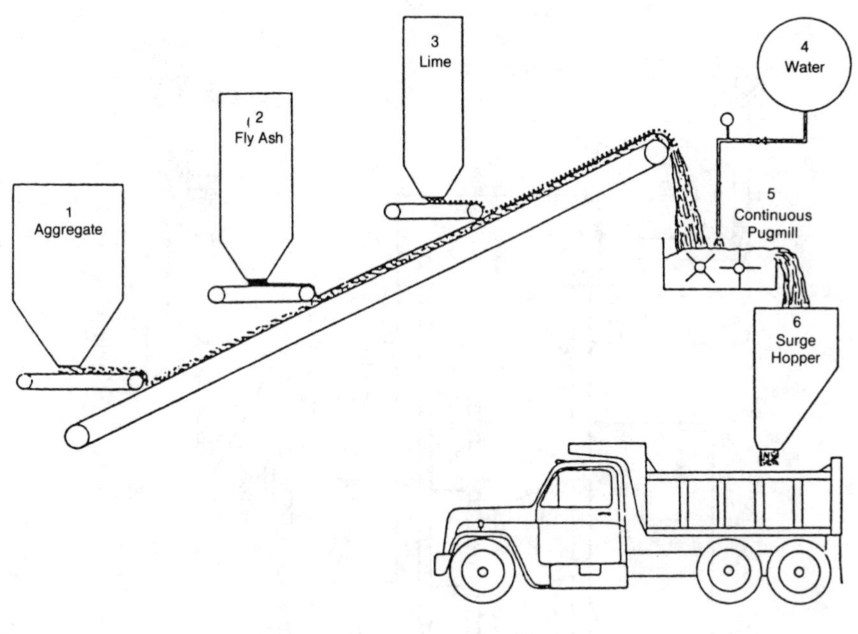

(Source: Transportation Research Board, 1976)

Figure 4.7 Typical lime—fly ash pozzolanic cement plant

Technical Feasibility of Solidification/Stabilization

Technical Description

Solidification processes include several types: stabilization, fixation, solidification and encapsulation and are described below:

- Solidification is defined as a process by which a sufficient quantity of solids is added to a liquid or semiliquid sludge to permit landfilling of the mixture with conventional earth-moving equipment, regardless of whether stabilization or fixation has occurred (Tittlebaum et al., 1985).
- Stabilization or fixation can be defined as the process of fully or partially bonding waste by adding to the soil a supporting medium, binder, or modifier. The modified waste that is produced should have improved physical and handling characteristics (Tittlebaum et al., 1985).

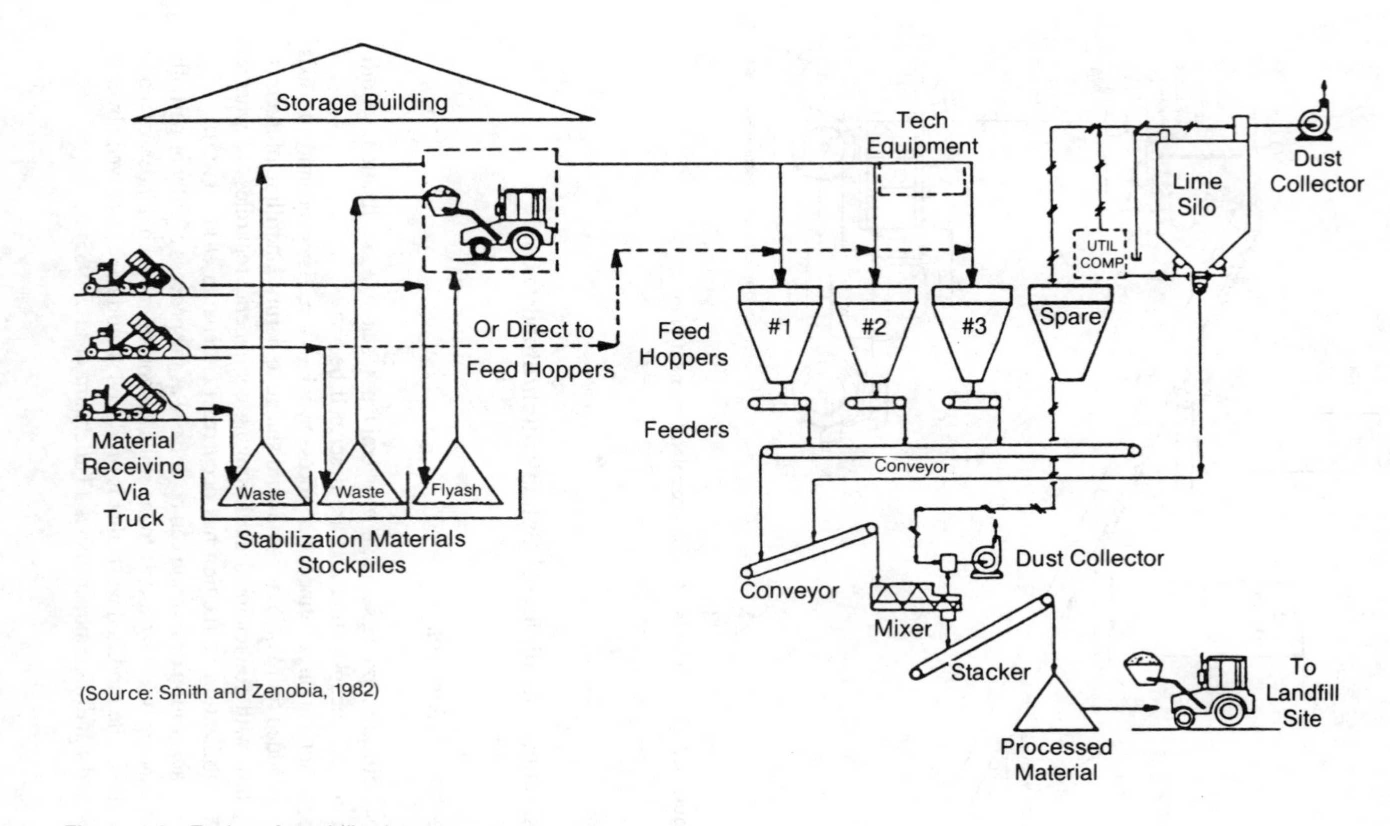

(Source: Smith and Zenobia, 1982)

Figure 4.8 Envirosafe stabilization process

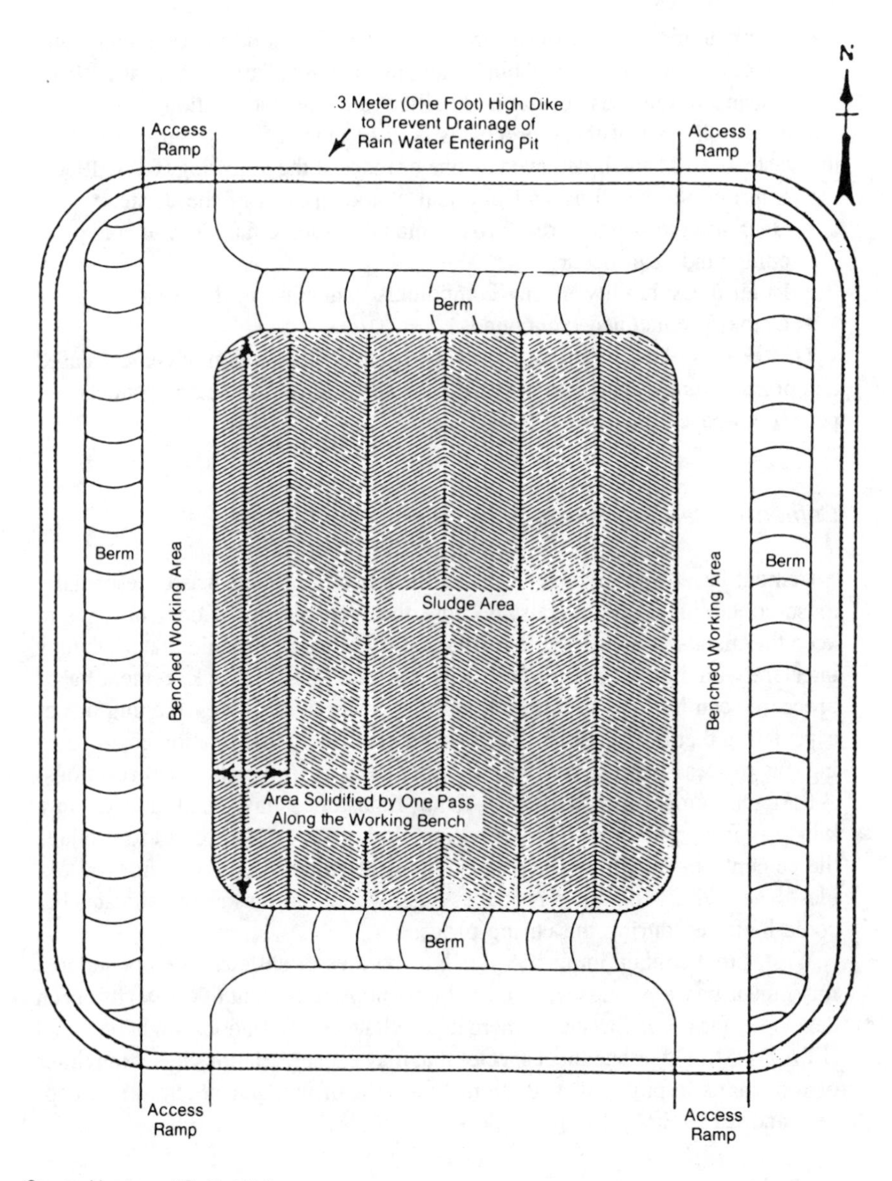

(Source: Musser and Smith, 1984)

Figure 4.9 In situ surface impoundment treatment

- Encapsulation is a physical process by which a solidification agent surrounds waste particles and binds them into a mass (Tittlebaum et al., 1985). Encapsulation thus results in stabilization, but not fixation.

The overall goal of these processes is either the safe disposal or reuse of the waste material and is achieved by one or more of the following (EPA, 1980):

- Improve the handling and physical characteristics of the waste.
- Decrease the surface area across which the transfer and loss of contained compounds can occur.
- Limit the solubility of any compounds contained in the waste.
- Detoxify contained compounds.

This section of the report describes the possible application of cement-based and pozzolanic processes to stabilize soils containing petroleum products and petroleum-based sludges.

Cement-Based Processes

Cement-based processes are most applicable to inorganic wastes, especially those containing heavy metals, because the high pH of the cement tends to keep the metals in an insoluble form. The presence of organic materials may interfere with the hardening of cements. Despite this drawback, cement-based processes can be used to limit the mobility of organics by entrapping or encapsulating them in the final solid mass. It is unlikely that fixation occurs during this process because organics do not take part in the chemical reactions.

Portland cement is prepared by firing limestone, clays, and silicates in a kiln at high temperatures. The resulting clinkers are then ground to produce the cement powder, a mixture of calcium, aluminum, silicon, and iron oxides. The addition of water produces a gel which hardens as the silicates become hydrated during the curing process.

Wastes that contain impurities, such as borates or sulfates, are not suitable for cement-based processes. A special low-alumina cement (Type V) has been developed for use in instances where high sulfate concentrations might be found (EPA, 1980). Other chemical species that may cause variations in setting times or reductions in physical strength include salts of manganese, tin, zinc, copper, and lead (EPA, 1980).

Pozzolanic or Lime-Based Processes

Pozzolan is defined as a siliceous or aluminous and siliceous material which in itself possesses little or no cementitious value. However, in finely divided form and in the presence of moisture, pozzolan will chemically react with calcium hydroxide at ordinary temperature to form compounds possessing cementitious properties (Mehta, 1983).

Fly ash, which contains silica and varying degrees of free lime, falls into the pozzolan category. Fly ash will form a gelatinous calcium silicate compound with cementitious properties when in the presence of water (Electric Power Research Institute, 1983).

Cement kiln dust is often used as a stabilization agent in pozzolan-based processes. Typically, cement kiln dust does not have the calcium content of portland cement. For this reason, it is sometimes referred to as a pozzolan, but it actually lies somewhere in between cement and pozzolans.

Pozzolan cements cure slowly, but have very high ultimate strengths and can be used effectively as a landfill construction material. The use of pozzolans applies mainly to massive concrete structures such as dams, spillways, roadway bases, etc. (Mather, 1969).

As with cement-based processes, most pozzolanic applications have been related to inorganic wastes. However, some organic wastes have been successfully stabilized. One industry representative claims that the only organic materials which interfere with the pozzolanic curing process are wastes containing sugars (C. L. Smith, Personal Communication).

Table 4.8 summarizes the advantages and disadvantages of cement and pozzolanic processes for the solidification of waste materials (EPA, 1980).

Experience

Several proprietary commercial pozzolanic processes for waste stabilization are available. These include the Calcilox process, developed by the Dravo Lime Company, and the Envirosafe process, originally developed by IU Conversion Systems for the specific purpose of converting waste products from coal burning power plants into landfill and construction material (Tittlebaum et al., 1985).

The Envirosafe pozzolanic process detailed in Figure 4.8 was used to solidify an oil-bearing sludge composed of about 37 percent by weight oil and grease (Smith and Longosky, 1984). Seven-day unconfined compressive strength exceeded 3.5 kg/cm^2 (50 psi) and permeability was less than 5×10^{-6} cm/s. Oil and grease content of the raw waste was 375,000 mg/kg (ppm). The oil and grease content of leachate samples from the solidified waste was reduced from 7,500 to 50 mg/kg (ppm).

Commercially available technologies utilizing cement-based processes include the Chemfix, Seolosafe, and Terra-Tite processes (Tittlebaum et al., 1985). All three processes are used primarily for inorganic waste, but small amounts of organics can be included with no adverse effects. Three documented

Table 4.8 Advantages and Disadvantages of Cements and Pozzolans

Advantages	Disadvantages
CEMENTS	
Inexpensive and plentiful raw materials.	Large quantities of cement often required.
High strengths and low permeabilities are possible.	Ultimate leachability not guaranteed.
Technology and equipment is commonplace.	Susceptible to borates, sulphates, acids.
System is tolerant of most chemical variations.	
Dewatering is not necessary.	
POZZOLANS	
Very inexpensive raw materials.	Handling of lime can be difficult.
High strengths and low permeabilities are possible.	Susceptible to acids, sugars.
Reactions are well understood.	Ultimate leachability not guaranteed.
Dewatering is not necessary.	

Source: EPA, 1980

cases involving the use of cement kiln dust to solidify oily wastes and sludges were also found.

In Dallas, Texas cement kiln dust was used in conjunction with the excavation and area mixing method described above to solidify 18,900,000 liters (5,000,000 gal) of 30-year-old oil sludge. Laboratory tests were performed as part of the engineering feasibility study and evaluated the use of clay, sulfur, cement, fly ash, cement kiln dust, stale cement kiln dust, quick lime, waste quick lime, sand, and crushed limestone in various combinations as solidification agents. A combination of stale and fresh cement kiln dust was found to be the best solidification agent available at a reasonable cost (Morgan et al., 1984). The on-site landfill area created by the solidification project provided suitable compressive strength to allow it to be developed as an industrial construction site.

Also in Texas, an abandoned oil field ''skim pit'' was solidified in situ by combining waste sludges with cement kiln dust at a ratio of one ton of kiln dust to one cubic yard of sludge (Musser and Smith, 1984). Class ''C'' fly ash was also evaluated for this project and appeared to provide slightly better load bearing values. However, the fly ash was rejected because the cement kiln dust seemed to absorb and encapsulate the oil more efficiently.

In Davenport, Iowa a 617.75 km^2 (2.5 acres) impoundment containing 25 years of accumulation of oily and PCB-bearing sludges, associated with the aluminum industry, was solidified in situ with the aid of cement kiln dust (Sonksen and Lease, 1982).

Implementation Feasibility of Solidification/Stabilization

Design Considerations

To achieve success in an operation of this sort, two difficulties must be addressed. First, a method must be devised to efficiently deliver the solidification/stabilization agent to the treatment area. Secondly, the sludge/agent mixture must be accomplished. The complexity of these two steps is largely a function of the site geography, depth of the sludge, and the sludge consistency. Specialized construction equipment may be needed to perform these tasks.

The choice of a solidification/stabilization agent is affected by many factors, including the physical and chemical characteristics of the material to be solidified and the purchase and transportation costs of the agent of choice. The initial choice of an agent should be made with the aid of a laboratory testing program designed to evaluate the effectiveness of a number of different agent/dosage combinations.

Equipment Requirements

Equipment requirements for solidification processes are highly dependent on site-and waste-specific factors. Solidification by the layering method described above uses common earth-moving equipment such as bulldozers, backhoes, and pulverizing compaction equipment.

In situ solidification of sludges contained in lagoons or other surface impoundments may require the use of specially designed or modified equipment. One of the two in situ solidification projects described above used long reach backhoes modified to include hydraulically operated injection systems and mixing augers. The second project used a pneumatic system to pump the cement kiln dust into the lagoon. This pneumatic system was coupled with a conventional backhoe bucket to perform the mixing.

Specific equipment needed to manufacture pozzolanic cements at either an on-or off-site facility is described in Figures 4.7 and 4.8.

Treatment Needs

Pretreatment is sometimes required for wastes which interfere with the curing of cements or pozzolans. Many of these pretreatment methods are proprietary in nature, but include the following types (Cullinane and Jones, 1985):

- Selected clays to absorb liquid and bind specific anions and cations.
- Emulsifiers or surfactants which allow the incorporation of liquid organic compounds.
- Other proprietary absorbents including carbon, zeolite minerals, or cellulose materials.

Disposal Needs

The solidification/stabilization section of this report has focused on the remediation of oily sludges found in surface impoundments. These processes may be applicable to oily soils as well. By utilizing a solidification/stabilization process to remediate a surface impoundment, the need to excavate and dispose of the oily soils that lie directly below the impoundment is eliminated.

Solidification/stabilization processes are often used in conjunction with some other remedial action such as a landfill liner or cover system in order to ensure that the final product will be secure. The ultimate use of the land on which the solidified material is disposed of is also an important factor. In most instances, extensive tests are conducted to determine both the leachability and the engineering properties of the solidified waste.

Monitoring Requirements

No standard tests exist to assess the success of a solidification/stabilization project (Tittlebaum et al., 1985). This is due largely to the great variety of wastes that are treated in this fashion. The tests that are performed fall under two broad headings of physical tests and chemical tests. Physical tests have the following objectives:

- Determine the particle size distribution, porosity, permeability, and wet and dry densities.

- Evaluate bulk properties.
- Predict the reactions of the material to applied stresses in embankments, landfills, etc.
- Evaluate durability.

Five standard tests that are often performed are bulk and dry unit weight, unconfined compressive strength, permeability, wet/dry durability, and freeze/thaw durability (Tittlebaum et al., 1985; EPA, 1980). These tests are documented in the Annual Book of ASTM Standards (1985).

Chemical tests are primarily concerned with evaluating the leachability of the solidified materials. There is a wide variety of leaching tests available in the literature. The EPA Extraction Procedure is considered to be the standard leachate test, although the ASTM Method A Extraction Procedure is often used. Briefly, the EPA Extraction Procedure exposes the solidified material to an acidic environment under the influence of stirring. After 24 hours, the filtered leachate is analyzed for a number of contaminants; primarily metals. This test was performed on leachate samples from the two in situ solidification projects described in this subsection. Trace metal concentrations in both cases were below the standards specified for wastes regulated under RCRA.

In addition to the above tests, monitoring wells may be required for surface impoundment solidification projects. Monitoring wells would be required to ensure that the lower sludge layers were adequately solidified and to track any plumes that may have developed prior to the waste solidification.

Permitting Requirements

Permitting requirements will vary from state to state. These processes are not currently considered to be standard disposal options for waste materials. The Environmental Protection Agency has published a "Guide to the Disposal of Chemically Stabilized and Solidified Waste" in an effort to clarify issues regarding this disposal option which states that under RCRA, solidification/stabilization options are treated only indirectly (EPA, 1980). This approach allows for inventiveness and flexibility as well as caution in granting of permits.

Environmental Feasibility of Solidification/Stabilization

Exposure Pathways

During excavation, installation, operation, and monitoring of a solidification system, possible primary pathways for direct human exposure include the following:

- Inhalation of vapors from the soil or groundwater.
- Particle inhalation and/or ingestion during excavation, installation, and operation when the waste is undergoing mixing or surface layering.
- Contact of skin with soil or groundwater during excavation and installation and with the treated streams during operations.

Possible primary pathways for environmental exposure during excavation and operation of a solidification system are groundwater and surface water. Surface water is also a primary pathway for environmental exposure during installation. Impacts to the groundwater and surface water often determine regulatory cleanup levels.

No secondary exposure pathways have been identified for this process (see Table 2.5).

Environmental Effectiveness

Solidification/stabilization processes have mainly been used in conjunction with inorganic pollutants. However, there are examples of successful solidification of oily wastes and sludges. Little information exists on the long-term nature, strength, and permanence of the bonds formed during solidification processes (Tittlebaum et al., 1985).

Two of the primary goals of solidification/stabilization processes are to decrease the surface area across which the transfer and loss of compounds occur and to limit the solubility of the contained compounds. Both of these goals are related to the leachability of the solidified materials. Solidification/stabilization processes are, therefore, designed to decrease the potential for exposure through groundwater pathways. In addition, potential exposure can be decreased and the durability of the solidified material can be enhanced through the use of a landfill liner or a capping system.

A third result is to detoxify contained compounds. When organic materials are solidified by cement or pozzolanic processes, it is unlikely that this occurs. Organic compounds do not react with the cement-like materials, but they are encapsulated so that they become unavailable to any exposure pathways. The toxicity of the contained compounds will then be less of an issue.

There are no quantitative measures of removal for solidification/stabilization processes. Success is measured primarily by the quality of leachate generated from laboratory or pilot tests and associated monitoring.

Economic Feasibility of Solidification/Stabilization*

Capital Costs

Capital costs involve the purchase of heavy equipment needed to perform the excavation and mixing operations. It is likely that this equipment would be leased or owned by a specialized contractor who would be hired to perform the work on a one-time basis. Equipment costs for impoundment excavation would be somewhat more expensive than for in situ solidification because this process requires more handling of the contaminated material and is also likely to be more land intensive.

Installation Costs

As is the case with capital costs, installation costs are likely to be somewhat higher for excavation of surface impoundments than for in situ solidification. Again, this is a function of the additional handling and land requirements associated with this process. On the other hand, it may be more difficult to achieve adequate mixing and solidification via the in situ approach. Total cost is, therefore, largely dependent on the quality of the equipment available, the experience of the contractor, and the nature of the waste to be solidified.

Specific cost breakdowns are not readily available in the literature. The surface impoundment excavation project described above resulted in total solidification and on-site disposal costs of $0.02 per liter ($0.08 per gal) or $21.12 per cubic meter ($16.16 per yd^3) of sludge (Morgan et al., 1984).

Table 4.9 provides estimated 1985 costs for silicate cement solidification of 1,890,000 liters (500,000 gal) or 2,590 metric tons (2,850 tons) of material by both the excavation/area mixing method and the in situ method. The material was assumed to be solidified by mixing with 30 percent portland cement and 2 percent sodium silicate. On-site disposal was assumed. Actual costs may vary significantly from those shown here because costs are highly waste- and site-specific. Excavation and area mixing methods were, in general, more expensive than in situ solidification for surface impoundments. Off-site solidification at specially designed plants were generally more expensive. However, the cost could be lowered somewhat if the waste were pumped into tank trucks for transport (Cullinane, 1985).

*All costs are presented in 1986 U.S. dollars unless otherwise specified.

Table 4.9 Summary of Efficiency, Processing Time, and Relative Cost of Solidification/Stabilization Alternatives

		Plant Mixing		
Parameter	**In Situ**	**Pumpable**	**Unpumpable**	**Area Mixing**
Metering and mixing efficiency:	Fair	Excellent	Excellent	Good
Processing days required:	4	1	14	10
COST ($ per metric ton):				
Reagent:	22.55 (63%)	22.55 (53%)	22.55 (42%)	22.55 (49%)
Labor and per diem:	1.50 (4%)	4.21 (10%)	7.62 (14%)	6.99 (15%)
Equipment rental:	1.52 (4%)	4.32 (10%)	8.20 (16%)	4.48 (10%)
Mobilization-demobilization:	1.74 (5%)	1.57 (4%)	2.49 (5%)	1.21 (3%)
Cost of treatment process:	27.31	32.65	40.86	35.23
Profit and overhead (30%)	8.19 (23%)	9.80 (23%)	12.26 (23%)	10.57 (23%)
TOTAL COST/TON	35.50	42.45	53.12	45.80

Source: Cullinane, 1985

Note: In all cases, 1,890,000 liters (500,000 gal) or 2,591 metric tons (2,850 tons) of waste treated with 30 percent portland cement and 2 percent sodium silicate with on-site disposal; costs include only those operations necessary for treatment. All costs are presented as dollars per metric ton ($/metric ton) of waste treated.

Operation and Maintenance Costs

While the solidification process is a one-time operation, there may be long-term operation and maintenance costs associated with either in situ or land-filled solidified wastes. Specifically, these may involve maintenance of cover systems or long-term sampling of monitoring wells.

Qualitative Ranking of Cost

In all of the examples cited above, solidification was chosen as the most economical of several alternatives, including landfilling, incineration, or soil recovery. Generally, solidification/stabilization processes are less expensive than the other technologies if the waste characteristics are compatible with the process. All of the examples discussed involved large volumes of semiliquid wastes and sludges which could not be disposed of in a conventional landfill. The oil content of the above mentioned wastes was high, but not high enough to allow for incineration or oil recovery. A further economic advantage results if the solidification agent can also be an industrial waste product, such as fly ash or cement kiln dust which can be acquired at a low cost.

GROUNDWATER EXTRACTION AND TREATMENT

General Description

When petroleum products that leak from underground storage tanks impact underlying groundwater, the remedial action chosen for that site usually includes measures designed to clean up the affected groundwater as well as the impacted soils. The various remedial measures available for impacted soils are discussed in the other subsections of this report; this subsection addresses those remedial actions associated with groundwater extraction and treatment. If the petroleum products reach the water table in sufficient quantity to form a floating layer, then recovery of the free-floating hydrocarbons (''product recovery'') may also be required.

Groundwater extraction and treatment is the process by which the affected groundwater is pumped out of the ground and treated by an appropriate treatment method to remove the compounds of concern. Once treated, the water is returned to the ground via injection wells or surface application. The treatment methods, as well as the feasibility of this approach, are discussed in detail in this subsection.

Experience has demonstrated that groundwater extraction and treatment is a very effective method for the removal of hydrocarbons from groundwater. However, this method is usually expensive and time consuming.

Process Description

All technologies discussed in this chapter share the need for an extraction well network to deliver groundwater to the treatment facility. Figure 4.10 schematically shows a typical pumping operation which is designed to prevent the continued migration of hydrocarbons from the area of the spill. Figure 4.10 also shows an injection well that can be used to recirculate the treated water

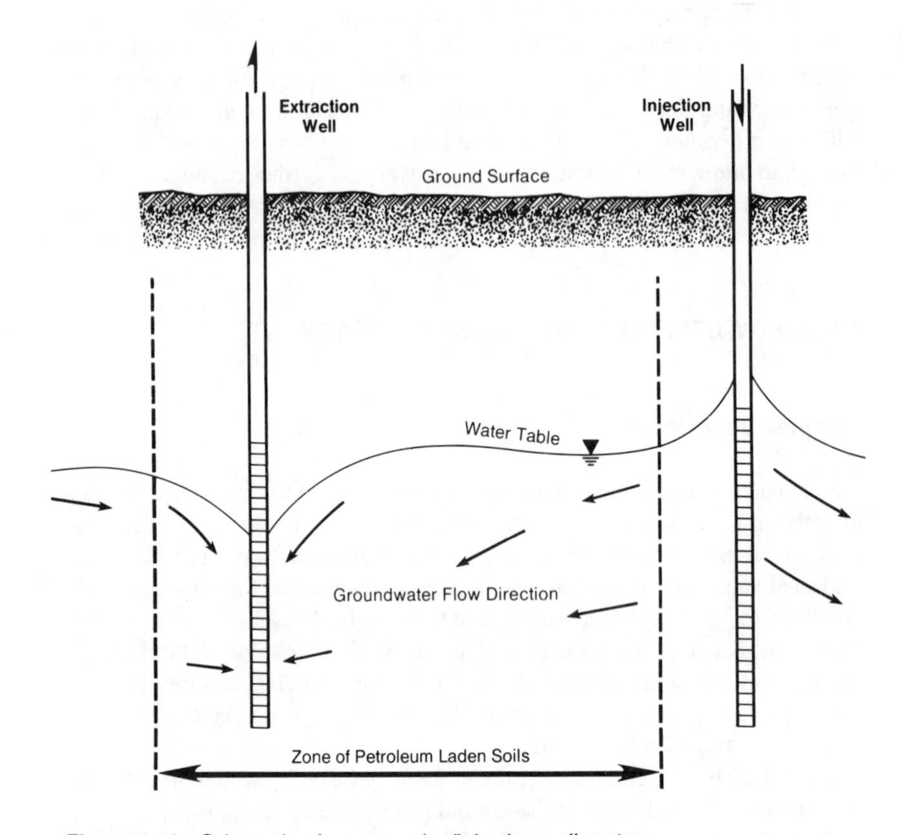

Figure 4.10 Schematic of an extraction/injection well system

back to the subsurface and enhance the hydraulic barrier created by the pumping wells.

Brief descriptions of the treatment system follow:

- Air stripping—Figure 4.11 depicts a typical air stripping operation in which groundwater is pumped to the top of a column filled with various packing materials such as tellerettes or Jaeger Tri-packs. Air is blown into the column counter-currently to the water flow and drives volatile organic compounds from the water.
- Oil/water separation—Figure 4.12 depicts one type of oil/water separator that utilizes closely spaced sinusoidally-shaped plates to collect oil floating to the surface of the water. Other filter-type separators are also available.

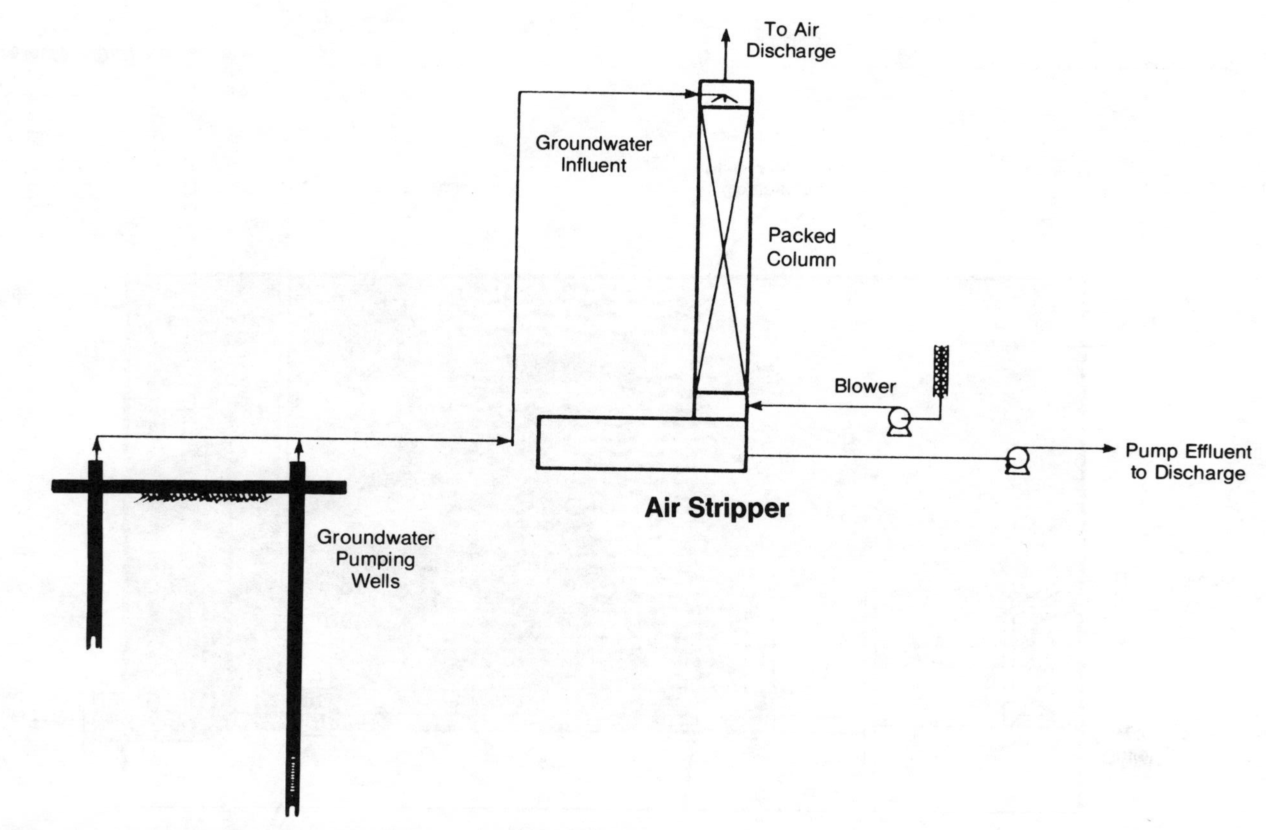

Figure 4.11 Typical air stripping installation for groundwater treatment

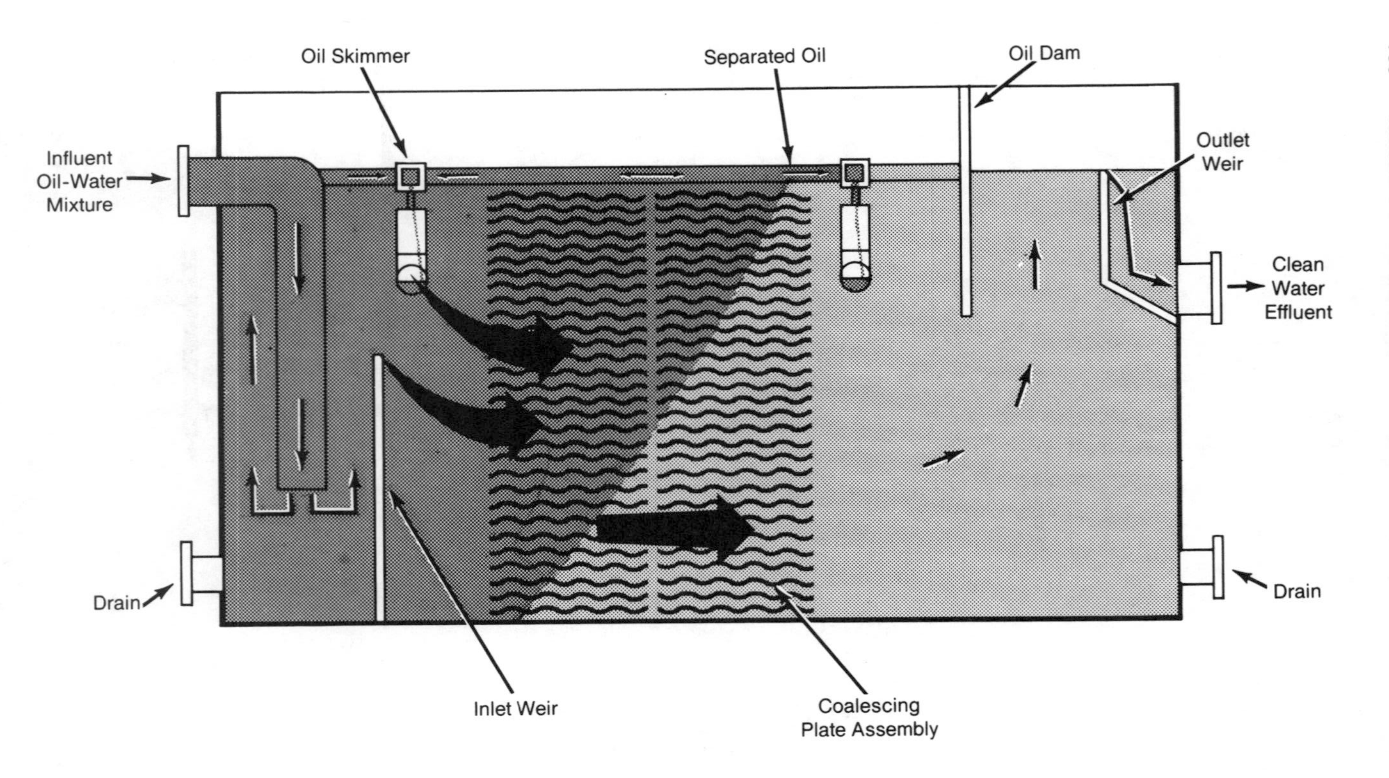

(Source: Fram Industrial, Undated)

Figure 4.12 Oil/water separator

- Carbon adsorption—Figure 4.13 depicts a typical carbon adsorption operation in which groundwater is pumped through multiple beds of granular activated carbon.
- Biological treatment—Figure 4.14 depicts a typical aboveground (non-in situ) biological treatment facility. The bioreactor intimately mixes microbial species, nutrients, and oxygen with the target hydrocarbons to enhance biodegradation.
- Product Recovery—Figure 4.15 depicts a two-pump scheme for recovering the oil layer floating on the watertable surface. Recovered hydrocarbon product may need little or no further treatment prior to reuse.

Technical Feasibility of Groundwater Extraction and Treatment

Technical Description

Groundwater extraction and treatment as a remedial action must address two primary issues. The first issue is the proper design of the extraction-injection well network. The second issue is the selection of the proper treatment method for the extracted groundwater.

The first issue can be addressed by determining the following site-specific factors which are prerequisities for adequately designing an optimum extraction/injection well system:

- Hydrogeologic characteristics (permeability, transmissivity, thickness of aquifer, depth of affected groundwater) will determine the depth, size, pumping capacities and separations between the extraction and injection wells.
- In addition to the above mentioned factors, water table level fluctuations with time are required to calculate the length of well screens and will provide information on groundwater flow paths and gradients. The latter are important for the proper design of well locations and spacings.

This information can also be used as input to a groundwater model that can assist in developing an optimum extraction/injection well system. Large sites with complex hydrogeology will require a model to simulate the designed network, whereas a small site with simple hydrogeology may not require the use of a model. Regardless of the size of the site and related conditions, ongoing monitoring will be required once the system is in place to assess the effectiveness of hydraulic control and related containment of the hydrocarbon compounds within a specific portion of the aquifer.

In addition to designing an optimum well network, the second issue is the selection of the proper treatment technology for the extracted groundwater.

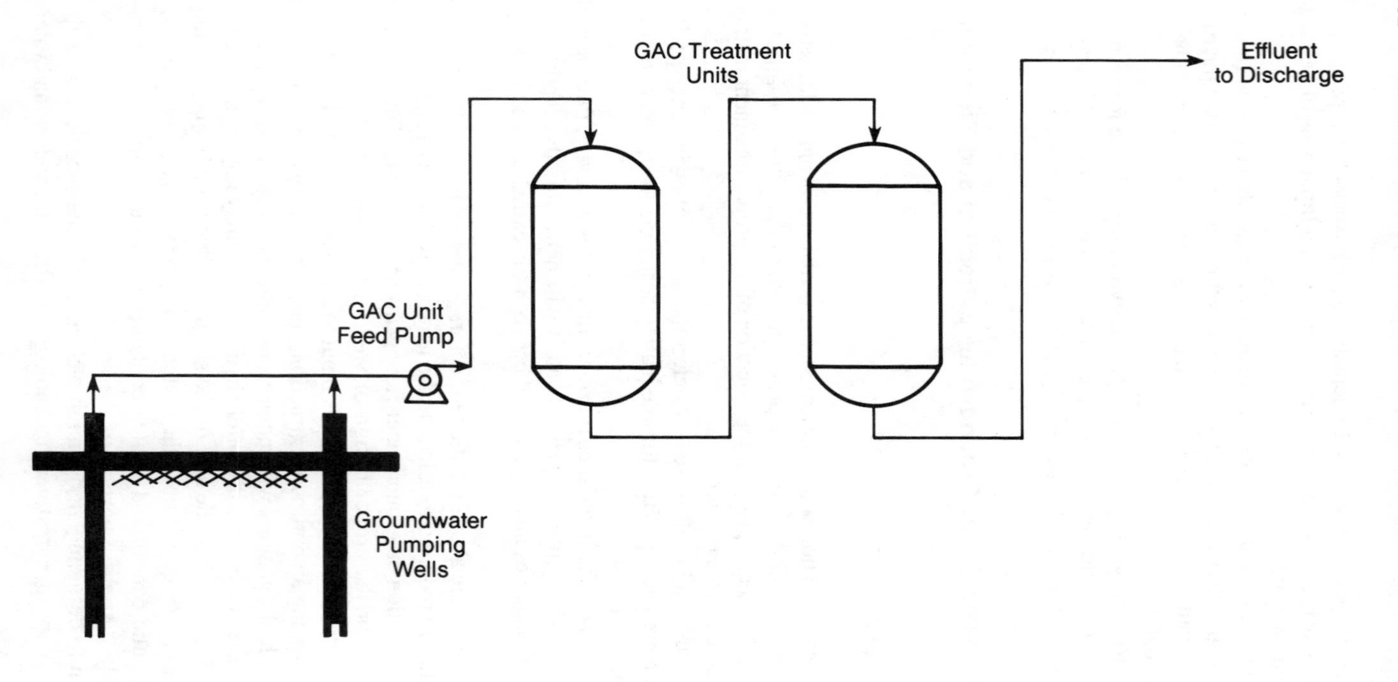

Figure 4.13 Typical granular activated carbon installation for groundwater treatment

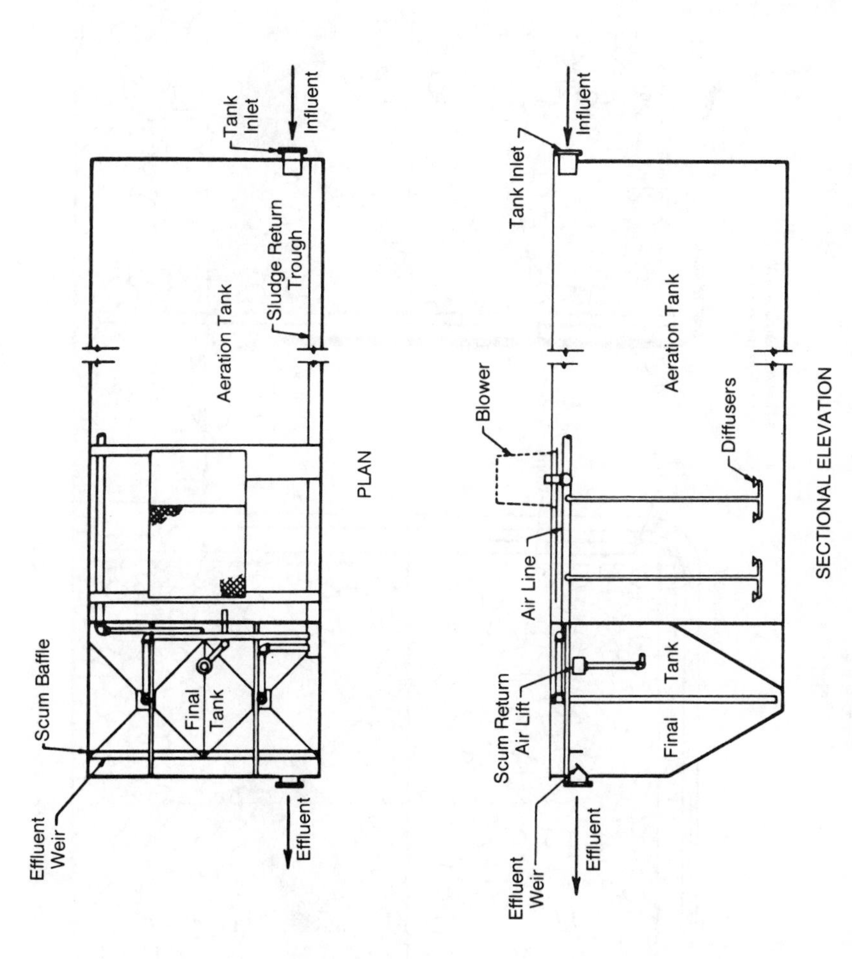

Figure 4.14 Extended aeration treatment plant for above ground biodegradation

(Source: EPA, 1977)

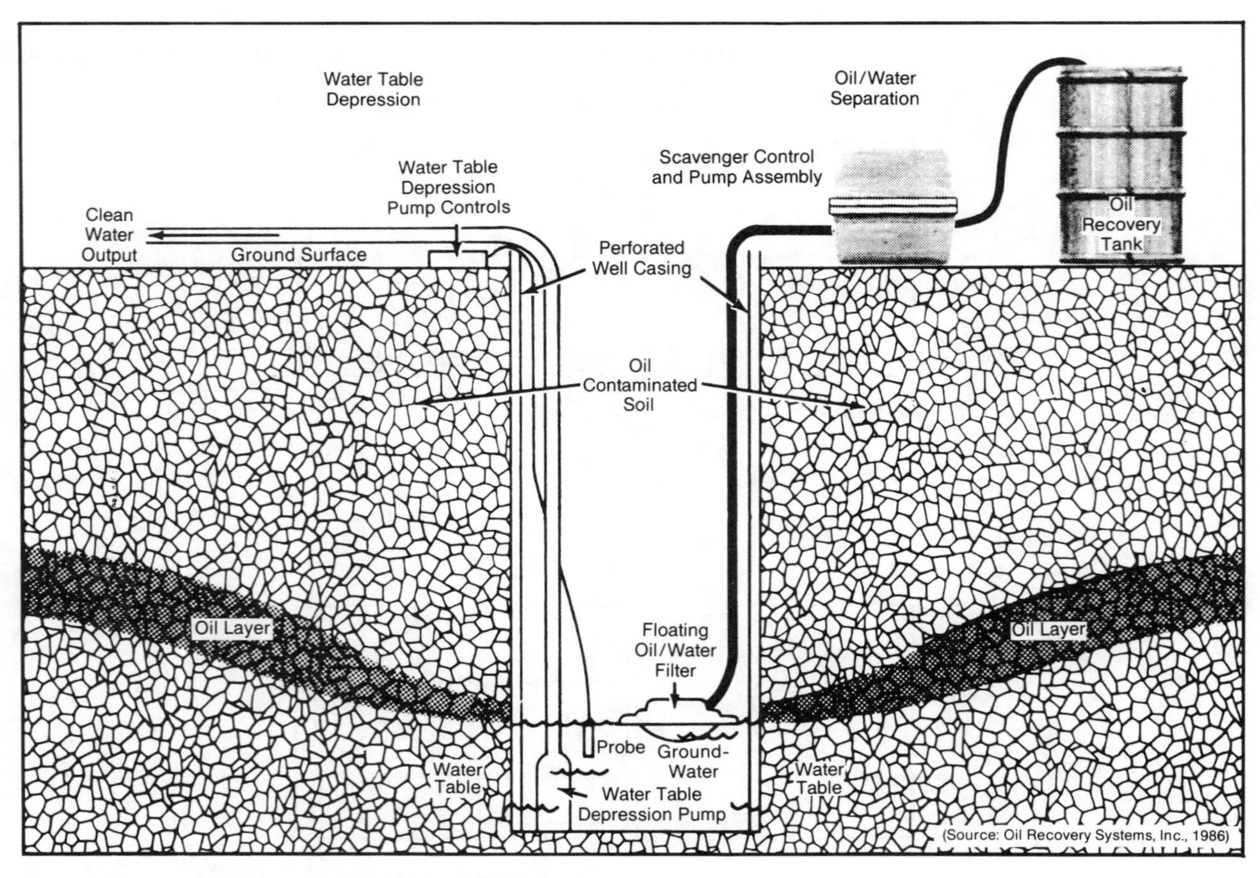

Figure 4.15 Product recovery using water table depression

Several technologies are available as outlined in Table 4.10. The appropriate treatment technique is chosen based on site-specific and chemical-specific factors as well as ultimate levels of treatment required or desired.

Table 4.10 Treatment Technologies for Groundwater Remedial Actions

Type of Treatment	Application Considerations
Physical Treatment Air Stripping	Air Stripping: Volatile compounds with large Henry's Law Constants (indicators of volatility) will be more readily air-stripped from treated groundwater than those compounds with low Henry's Law Constants (Kavanaugh and Trussell, 1980).
Oil and Water Separation	Oil/water Separation: Used to remove large quantities of immiscible petroleum products from groundwater. This method is applicable if the petroleum products and water exist as separate or emulsified phases.
Biological Treatment	Biodegradation: Above ground biological reactors can be used to treat the pumped groundwater and utilize similar mechanisms outlined for in situ biodegradation (EPA, 1977).
Chemical Treatment Carbon Adsorption	Granular Activated Carbon (GAC): When present at low levels, some of the water-soluble organic components of gasoline and other products can be more effectively removed by adsorption onto granular activated carbon (GAC) than by air. GAC is often used as a secondary step in groundwater treatment. Other technologies, such as air stripping, biodegradation and oil/water separations, are used initially to remove large amounts of petroleum products from the groundwater. Treatment with GAC then follows as a secondary step to "polish" the treated stream in order to reduce the concentrations to acceptable target levels.

Notes:

1. These treatment technologies all use a groundwater extraction/injection system to transport the groundwater to the surface for treatment and/or control of the groundwater gradient and resulting flow direction at the site.

2. "Dual Pump Hydrocarbon Recovery" refers to the technology which uses two pumps within one well. One pump, or skimmer, is installed to recover the floating layer of petroleum product. The second pump is installed deeper in the well and is used to create a cone of depression that controls the hydraulic gradient. If the petroleum product is present in sufficient quantities, it can be refined and reused after removal. The groundwater is then treated using one of the other technologies.

Experience

Case studies describing two of the more common pump and treat technologies, carbon adsorption and air stripping, are presented in this section. Although these examples refer to specific compounds, the principles are applicable to a variety of organic components.

Case Studies for Carbon Adsorption. In early 1980, the Township of Rockaway, New Jersey discovered trichloroethylene (TCE) in two of its three public wells which served 3,000 families. The U.S. Environmental Protection Agency (EPA) recommended no more than 4.5 micrograms per liter (μg/L) or parts per billion (ppb) TCE in this drinking water supply (Jacobson, 1981). Early testing showed consistently high levels of 50 to 100 μg/L (ppb) in one well and 170–220 μg/L (ppb) in the other. Both of these wells were taken out of service. The third uncontaminated well continued to supply the needs of the Township.

In October 1980, Township officials responded to complaints about taste and odor in the drinking water and ordered a chemical analysis performed. Two additional chlorinated hydrocarbons were discovered, diisopropyl ether and methylene chloride, in concentrations of 80 to 85 μg/L (ppb) and 40 to 45 μg/L (ppb), respectively.

A water emergency was declared and an activated carbon filtration system installed. The treatment system incorporated two adsorption tanks, 3.05 meters (10 ft) in diameter and 6.1 meters (20 ft) high, each holding 9,076 kg (2,000 lbs) of granular activated carbon. System capacity was 3,780,000 liters (1,000,000 gal) per day and removed the target organics to concentrations of less than one microgram per liter which was below the recommended action level. The quantity of contaminanted water that could be treated before replacing or regenerating the activated carbon was not available for this study.

While this system was designed for removal of chlorinated organics, a very similar system could have been installed to remove low concentrations of petroleum hydrocarbons. For example, granular activated carbon was used in Wausau, Wisconsin to treat drinking water that had been affected by a spill from an underground gasoline storage tank (Hall, 1985). The resulting treated water had undetectable concentrations of benzene, toluene, and xylenes.

Case Study—Air Stripping. In 1981, the City of Wausau, Wisconsin discovered that water from some of its wells contained volatile petroleum hydrocarbons (Hand et al., 1986). These wells were a primary source of water for the city. The water contained an average of 30.9 μg/L (ppb) toluene, 16.6 μg/L (ppb) total xylenes, and 5.1 μg/L (ppb) ethylbenzene. In October 1983, a project was undertaken by the U.S. Environmental Protection Agency (EPA), the City of Wausau, and Michigan Technological University to investigate the cost effectiveness of using a packed tower air stripper to remove the VOCs from the groundwater.

The study resulted in the design of a full-scale air stripping tower with design parameters, as presented in Table 4.11. The 7.60 meter (3-in.) plastic saddle packing for the air stripper was chosen over the 5.1 cm (2-in.) Tri-packs for the following reasons:

- Iron precipitation, which can cause tower clogging was minimized by the use of saddles which have a smaller area on which precipitates could form.
- Saddles were less expensive.
- The tower volume was greater with saddles allowing for replacement with Tri-packs if greater removals became necessary and iron precipitation did not result in a problem.

During the first four months of operation, no major problems were encountered. The tower processed 5,670 liters (1,500 gal) per minute and reduced the average concentrations of toluene, xylenes, and ethylbenzene in the effluent to 0.94 μg/L (ppb), 0.60 μg/L (ppb), and less than 0.3 μg/L (ppb), respectively. The system is still in service.

Table 4.11 Design Parameters for the Full-Scale Air Stripping Tower

Parameter	Value
Air-to-water ratio	30
Tower pressure drop	52N/m^2/m (0.06 in. H_2O/ft)
VOC removal (based on TCE)	95 percent
Henry's Law constant	0.116
Packing type	0.0762 m (3 in.) plastic saddles
Water flow rate	0.0964m^3/s (1500 gpm)
Air flow rate	2.84 m^3/s (6000 cfm)
Water loading rate	20.1 kg/m^{2}s (29.8 gpm/sq ft)
Air loading rate	0.755 kg/m^{2}s (119 cfm/sq ft)
Temperature	10°C (50°F)
Tower diameter	2.44m (8 ft)
Tower height	7.47m (24.5 ft)

Source: Hand et al., 1986

Implementation Feasibility of Groundwater Extraction and Treatment

Design Considerations

Remedial options have, as their goals, containment of the hydrocarbon plume and removal of hydrocarbons from the groundwater. These options may require extensive extraction well networks to contain the migration of hydrocarbons and depend upon the site hydrogeology and size of affected area.

The treatment option which is chosen should depend principally upon the chemical constituents present in the groundwater, their concentrations, and the cost effectiveness of the method. While the pumping rate is heavily dependent upon site-specific hydrogeology and cannot be entirely predicted for a certain site, the treatment options described above can be anticipated for certain chemical constituents. Table 4.12 presents these chemical groupings and applicable treatment technologies.

Equipment Requirements

All extraction/treatment options share the basic well pumping system. The treatment equipment requirements will vary with the treatment option chosen.

Table 4.12 Type of Petroleum Product and Applicable Technology for Groundwater Treatment

Applicable Treatment Technology	Type of Petroleum Product/Constituent
Air Stripping	Volatile Organic Compounds (VOCs): benzene, toluene, lighter gasoline-like products
Oil/Water Separators	All petroleum products in immiscible phase
Carbon Adsorption	Aromatics, higher molecular weight lower solubility constituents
Biodegradation	Lower molecular weight, higher solubility constituents
Hydrocarbon Recovery	All petroleum products in lighter-than-water immiscible phase

The major pieces of equipment for each treatment option are noted below:

- Air stripping utilizes a packed bed tower which is typically 0.3 to 1.22 meters (1 to 4 ft) in diameter, as much as 6.1 to 9.1 meters (20 to 30 ft) in height, constructed of fiber-reinforced plastic, packed with polypropylene packing and equipped with an explosion-proof blower.
- Oil/water separators are available in a variety of designs. Different designs will have different hydrocarbon removal efficiencies. One design utilizes closely packed plates to enhance the gravity-driven separator of oil and water, whereas other designs preferentially adsorb the hydrocarbon fraction.
- Carbon adsorption utilizes drums or tanks (often called "beds") packed with activated carbon. These beds are usually in series with monitoring between beds to detect for carbon exhaustion in the lead bed.

Several factors determine the quantity of carbon needed to treat a certain volume of water containing a given concentration of specific organic contaminants. Once the capacity of the carbon has been reached, the used carbon must be replaced with clean carbon. Some typical ranges of design capacities are: 0.05 to .45 kg (0.1 to 1 lb) of carbon per 3,780 liters (1,000 gal) of water treated having concentrations in the μg/L (ppb) contaminant; and 0.9 to 9 kg (2 to 20 lb) of carbon per 3,780 liters (1,000 gal) of water having concentrations in the ppm range.

- Biodegradation utilizes various designs to provide intimate contact of microbes, oxygen, nutrients and target hydrocarbons. Large aeration tanks and ponds or bioreactors with air spargers are used. With typical 24-hour retention times for the treated groundwater, these units are much larger for the same flow rate than units which utilize the other treatment technologies.
- Water table depression/hydrocarbon recovery utilizes a dual pump system to draw the floating hydrocarbon to the well with one pump and to remove it from the well with a second.

The technologies described above can be utilized in the following ways:

- An existing on-site wastewater treatment plant can be used. However, such a facility may not be suitable for removal of trace concentrations of petroleum products.
- A mobile system can be used for pretreatment followed by discharge to an existing on-site wastewater treatment plant (if the plant is suitable for target treatment levels).
- A mobile or on-site system can be used with effluent discharged to either surface or groundwaters if treated levels correspond to target levels.

Treatment Needs

Technologies for groundwater extraction and treatment are chosen to address the treatment needs of specific site conditions and regulatory policies. A common approach is to combine technologies to achieve effective treatment and meet discharge criteria. Air stripping, for example, is often combined with carbon adsorption or product recovery to enhance its optimal operating range. Table 4.13 highlights the hydrocarbon removal efficiency that can be expected for each option.

Disposal Needs

The production of by-products or end-products requiring disposal is often the case in remedial activities. All of the technologies discussed in this section for groundwater treatment generate a water effluent stream. This effluent must be of sufficient quality to meet surface discharge criteria. The following disposal requirements may also be necessary for the technologies under discussion:

- Air Stripping—Under a variety of circumstances an air stripper is subject to blockage by scale buildup due, in part, to water hardness and/or certain forms of bacteria. This scale buildup along with the packing attached to it, requires disposal. Air discharge from the stack may require treat-

Table 4.13 Hydrocarbon Removal Efficiency for Various Groundwater Treatment Technologies

Treatment Technology	Hydrocarbon Removal Efficiency
Air Stripping	95 to 99 percent removal of influent VOCs is possible depending on composition of VOCs.
Carbon Adsorption	99 percent removal is possible for adsorbing constituents.
Oil/Water Separators	Effluent oil concentrations below 15 mg/L (ppm) are possible.
Biodegradation	Final effluent concentrations of bio-utilized constituents ranging from 100 to 1,000 μg/L (ppb) are possible.
Hydrocarbon Recovery	Recoverable hydrocarbons can be pumped from oil slicks as thin as 0.5 cm (0.2 in.).

ment; options include carbon adsorption or thermal treatment. Should vapor phase carbon be required to treat the tower stack effluent, disposal or regeneration of the spent carbon will be required.

- Carbon Adsorption—Spent carbon can be sold, regenerated or disposed.
- Oil/Water Separation—When filter-type separators are used, spent cartridges will require disposal. Plate-type devices will separate out solids, if any, which require disposal.
- Biodegradation—Above ground bioreactors may develop excessive quantities of biomass and settled solids, requiring clean-out and disposal. These by-products may require dewatering prior to disposal.
- Product Recovery—The pumped groundwater from a two-pump system may require further treatment using one of the above methods and/or disposal.

Monitoring Requirements

All of these technologies share the requirement of monitoring for continued migration of hydrocarbons and monitoring for effective remediation. The former can be readily accomplished by appropriate placement of groundwater monitoring wells. Samples of pumped groundwater taken prior to and after treatment, will not only document declining concentrations of target hydrocarbon compounds but also efficiencies and ultimate effectiveness of the treatment. Additionally, vapor effluent from air strippers may be monitored for hydrocarbon release to the atmosphere, carbon beds may be monitored for carbon exhaustion, and bioreactors may be tested for microbial population.

Permitting Requirements

Permits from several government and regulatory agencies may be required. On a local level, establishing any treatment system may require compliance with building and land-use ordinances. The National Ambient Air Quality Standards through the Clean Air Act establishes guidelines enforceable through state or Federal agencies governing the air emissions from any treatment system (primarily air stripping systems). Treated water discharged to surface streams at a point source discharge may require compliance with National Pollutant Discharge Elimination System (NPDES) standards. Discharge to an existing treatment plant may require permit modifications to handle the additional flow/loading. Reinjection of treated water to the subsurface may require a state or local permit, similar to the NPDES permit.

Environmental Feasibility of Groundwater Extraction and Treatment

Exposure Pathways

Extraction and treatment technologies, separately or in combination, can remediate the water-soluble constituents of the spilled petroleum product (i.e., the aromatics such as benzene and phenols). While these constituents are hazardous, the treatment options discussed here can be designed and operated to minimize potential exposure to personnel.

For extraction and treatment technologies, the possible primary pathways for direct human exposure during excavation, installation, operation, and monitoring include the following:

- Inhalation of vapors from the soil or groundwater during excavation, installation, and monitoring and/or inhalation of emissions from the treatment unit during operations and monitoring.
- Particle inhalation or ingestion during excavation and installation.
- Contact of skin with soil or groundwater during excavation, installation, and monitoring and with the treated stream during operations and monitoring.

The primary environmental exposure pathways for extraction and treatment technologies include groundwater, surface water, and air. Often, the extent to which the waters are affected dictates the regulatory cleanup levels.

Secondary pathways include particle inhalation and ingestion during operation and monitoring when these impacts are minimized.

Environmental Effectiveness

The use of these technologies, often in combination with one another, has been effective at restoring groundwater quality. Restoration can occur over weeks or years, depending upon the extent of hydrocarbon migration, cleanup levels required and other chemical and hydrogeological factors. Residual hydrocarbon concentrations in the low part-per-billion to nondetectable range are possible depending on the compounds and the method of treatment.

Economic Feasibility of Groundwater Extraction and Treatment*

Capital Costs

Capital costs vary with the treatment scheme presented and the size of the unit. The number, depth, and type of pumping (and injection) wells may be a factor. A mid-sized well capable of handling a dual pump recovery system (i.e., 0.15 meter (6 in.) PVC) can typically cost $32.81 to $65.62 per linear meter ($10 to $20 per linear foot) for materials only. Well pumps, dual recovery pumps, and product recovery oil-filtering pumps can cost several hundred dollars each.

The capital costs for carbon adsorption, air stripping and aboveground biodegradation are all a direct function of flow rate. Influent concentration will also affect treatment facility designs and therefore capital costs. The smallest practical air stripping tower of one foot diameter can handle flow rates less than 76 to 95 liters per minute (20 to 25 gpm) and cost less than $20,000. A comparably sized carbon adsorption unit will cost approximately $5,000, but may require periodic purchase of new activated carbon. The cost of a bioreactor or aeration basin can be significantly greater than the other methods due to the size of such units.

The cost effectiveness of groundwater treatment systems is often time-dependent. An air stripping treatment system, for example, generally has greater initial capital costs than a carbon adsorption treatment system. However, the generally higher operating and maintenance costs of carbon adsorption make air stripping more economical for prolonged operations. Figure 4.16 graphically presents this relationship.

Installation Costs

All treatment options share the $98.43 to $164.04 per linear meter ($30 to $50 per linear foot) installation cost for a mid-sized well (i.e., .15 meter (6 in.) PVC), with the possible exception of biological treatment. The installation costs for all options listed here are minimal since all require very little site preparation and can arrive in skid-mounted packages.

*All costs are presented in 1986 U.S. dollars unless otherwise specified.

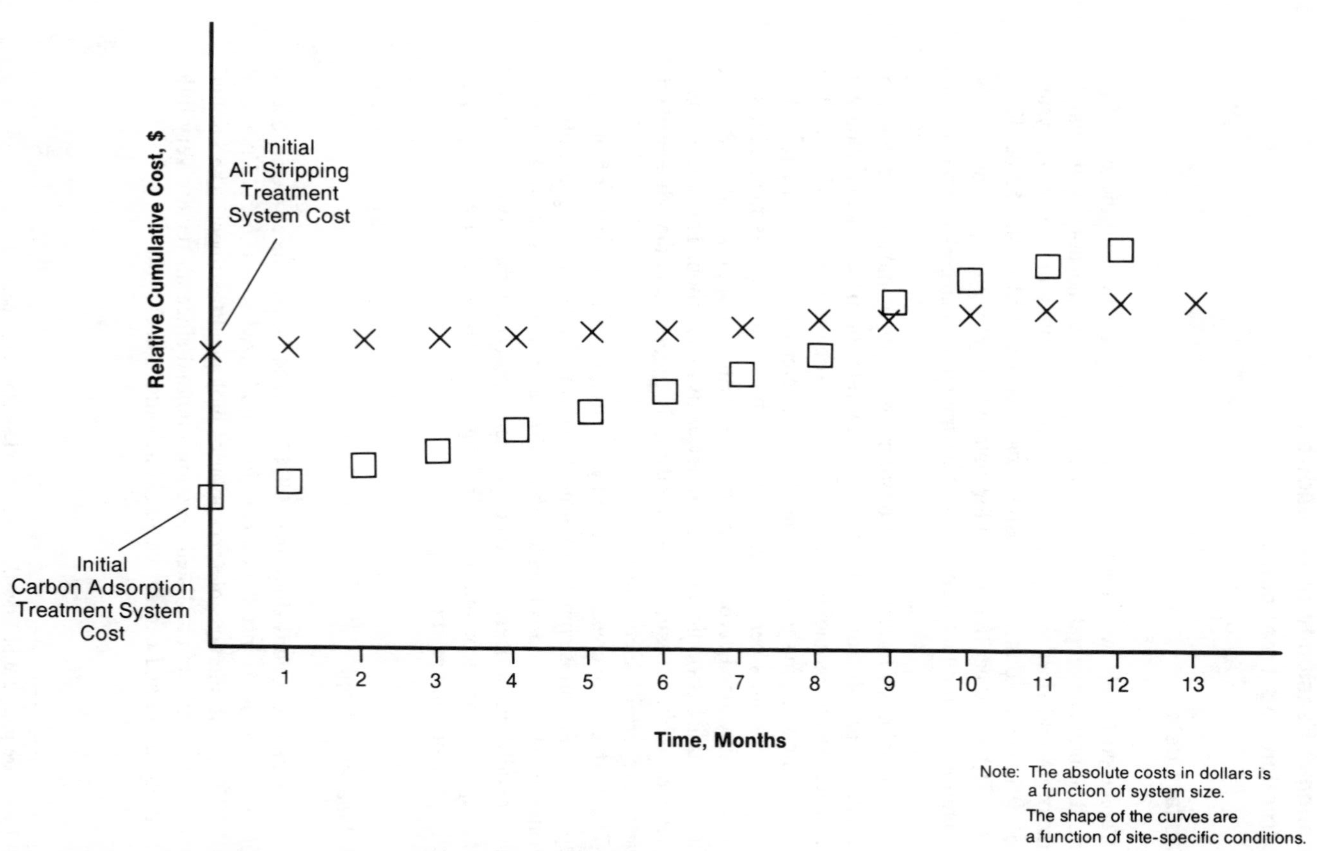

Figure 4.16 Air stripping and carbon adsorption treatment system costs

Operation and Maintenance Costs

Both air stripping and biodegradation share the potential problem of accumulation/blockage by settled solids, scaling and biomass. Periodic clean-out of these units should be expected. In the case of a seriously plugged tower, replacement of the packing material may be required at $714.3 to $1,785.7/cubic meter ($20 to 50/ft^3). A small tower may have 0.56 cubic meters (20 ft^3).

Periodic replacement of activated carbon should be expected. Carbon lifetime is a function of the flow rate, chemical constituent being treated, and influent concentration. Costs range from $0.05 to $0.66/1,000 liters ($0.20 to $2.50/1,000 gal) (O'Brien and Fisher, 1983).

Oil/water separation based on filtration principles will require periodic cartridge replacement and disposal. Infrequent replacement of small cartridges will have minimal costs. Product recovery operations have only minor O&M costs and may actually contribute a small source of revenue from recovery of the product.

Qualitative Ranking of Cost

Of the technologies discussed in this section, product recovery and oil/water separation rank low in initial cost and operation. The remaining technologies rank low to moderate in cost and can be expensive in the case of very high pumping rates (and corresponding large treated volumes). The extraction well system can also vary widely depending on the size of the impacted area.

CHEMICAL EXTRACTION

General Description

Chemical extraction is the process by which excavated soils are washed to remove the compounds of concern. This is typically accomplished by the utilization of a washing plant which utilizes a water/solvent or a water/surfactant mixture to remove the compounds. This method is very similar to the in-situ leaching process described in Section 3. The primary difference is that by removing the soil from the ground, wash mixtures can be utilized which increase product recovery. These mixtures cannot be used in situ because they would adversely affect the environment. This process, as well as its feasibility, is discussed in more detail in this subsection.

Experience has shown that chemical extraction is an effective method for the removal of hydrocarbon compounds from the soil. However, the costs of a treatment plant and the extraction mixture are still relatively high.

Process Description

The chemical extraction process proceeds as follows:

- Excavated soil is passed over a wide mesh screen for size reduction and removal of large objects.
- Solvent/surfactant is added to the soil and intensively mixed.
- The solvent/surfactant and extracted hydrocarbons are separated from the soil.
- The soil may be washed or subjected to aeration for removal of the extracting agent/solvent.
- The extracting agent (a surfactant or solvent) is filtered for fine particles.
- The extracting agent is treated to remove the hydrocarbons.

Figure 4.17 schematically depicts a chemical extraction operation.

Technical Feasibility of Chemical Extraction

Technical Description

Chemical extraction of petroleum hydrocarbons from excavated soil can be accomplished by washing the soil in a suitable solvent or surfactant. The choice of a solvent is dependent on the solubility of the target compounds in that solvent. Polar solvents, such as water, tend to cause only polar compounds to go into solution. Most petroleum hydrocarbons are nonpolar in nature. Therefore, water is not a very effective extraction agent for these compounds.

Surfactants and organic solvents are more effective than water as extraction agents for petroleum hydrocarbon compounds. Organic solvents are nonpolar and, as a result, they have the capability to dissolve nonpolar hydrocarbon compounds. In contrast to organic solvents, surfactants, such as soap, have both polar and nonpolar ends and can dissolve nonpolar hydrocarbons while maintaining their own solubility in water (a polar solvent).

The degree to which chemical compounds are adsorbed onto soil particles also has a significant effect on the potential effectiveness of the extraction process. As a result, the degree to which these hydrocarbons are dissolved in the solvent and recovered depends on the relative affinity (or partitioning) of the hydrocarbons to the soil and solvent.

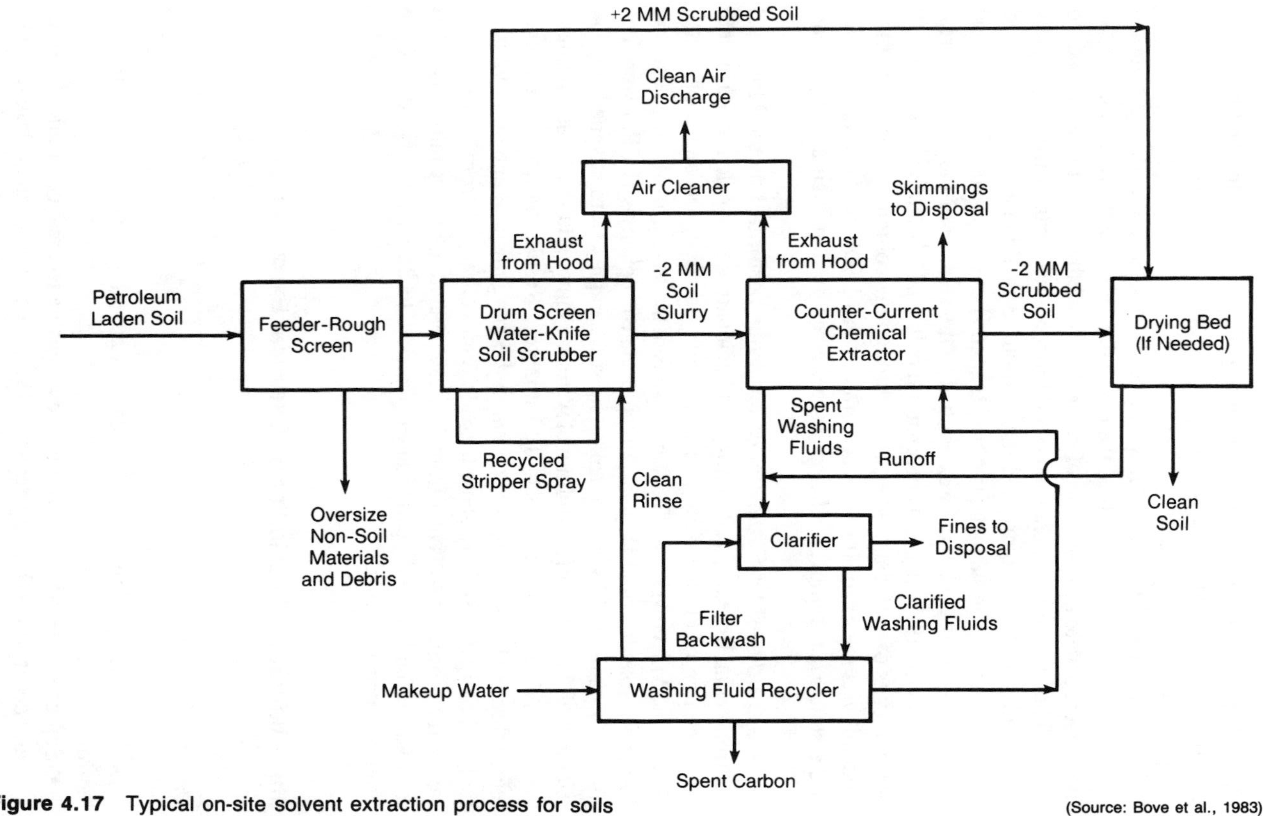

Figure 4.17 Typical on-site solvent extraction process for soils (Source: Bove et al., 1983)

Experience

Experience in chemical extraction from soils is very limited and restricted to pilot- and bench-scale facilities.

Case Study—Bench-Scale. Work completed at the Volk Air National Guard Base, Wisconsin involved the extraction of crude oil from soils (Nash and Traver, undated). A mixture of crude oil and methylene chloride was added to soil. This mixture was chosen because it contained many different organic compounds. Soil was compacted into 7.6 cm (3 in.) diameter, 1.5 meter (5 foot) high tubes. The surfactant solution, a blend of ethoxylated fatty acid and an ethoxylated alkyl phenol, was applied to the surface of the soil at a pressure of 50 cm (19.7 in.). Gas chomatographic analysis showed that after 10 pore volumes of surfactant was passed through the column, 88 percent of the crude oil was removed. Initial concentration was 1,000 mg/L (ppm) organics in the soil. Final removal efficiencies were reported as high as 98 percent.

Case Study—Pilot-Scale. Based on testing performed by the Municipal Environmental Research Laboratory in Cincinnati, a full-scale mobile system for extraction of spilled materials from soils was designed and constructed (Scholz and Milanowski, 1983). Some results of the testing are discussed below.

Phenol, arsenic trioxide, and polychlorinated biphenyls (PCBs) were applied to organic and inorganic soil. The soil samples were stripped for one minute with a water knife. Additional water washing of the soil was continued for various time intervals up to 120 minutes. Results indicate that phenol removal ranged from 97 to 99 percent in the inorganic soil and 60 to 92 percent in the organic soil. Arsenic trioxide removal ranged from 28 to 52 percent in the inorganic soils and from 47 to 59 percent in the organic soils. PCB removal in the inorganic soils ranged from 21 to 28 percent. No data were listed for organic soils.

Implementation Feasibility of Chemical Extraction

Design Considerations

Design considerations for the chemical extraction process should include the following:

- Soil characteristics—Soils must be amenable to breakdown, dewatering, and desorption (Huibregste et al., 1980). The clay content should be less than 20 to 30 percent, as clays tend to have a significant adsorption capacity. Also, soils with high organic content tend to adsorb more hydrocarbon compounds than inorganic soils. The degree of adsorption will be a contributing factor in the evaluation of the degree of difficulty to be encountered for the process.

- Hydrocarbon characteristics—The chemical nature of the petroleum hydrocarbon, particularly its miscibility with the extraction solvent, is critical to the operation.
- Solvent characteristics—Extraction of aqueous hydrocarbons is usually achieved by using aqueous solutions of HCl, NaOH, Na_2CO_3, surfactants or organic solvents. The solvent should have a low volatility to prevent air contamination and solvent losses. Potential solvents should have low toxicity since residual concentration may remain on the "washed" soils. The use of organic solvents is not widespread; however, a proprietary solvent believed to be a sodium polyethylene glycol mixture is currently being tested (Beaudet et al., 1983). The separation of the solvent from the petroleum hydrocarbons for recycle is important for economical feasibility (Scholz and Milanowski, 1983).
- Soil water content—Water initially present in the soil limits the extent of petroleum hydrocarbon adsorption onto soil particles.
- Elapsed time from spill—The length of time that the hydrocarbons have been in the soil directly affects the extent to which adsorption takes place. As time passes after the initial spill, hydrocarbons continue to adsorb onto internal surfaces of soil particles making extraction more difficult.

The decision to utilize a transportable unit or construct a permanent installation is often dictated by cost considerations. The cost of the entire cleanup by both methods should be evaluated. A permanent facility will be very costly to establish (see Implementation Feasibility). However, once in place the operating costs can be modest compared to the operating costs and fees associated with a leased portable unit. Generally, it would be more cost efficient to process smaller quantities of soil using a leased unit.

Equipment Requirements

Equipment requirements vary according to site specific conditions. Generally, equipment needed for the process includes the following (Smith, 1985):

- Pretreatment of soils—Equipment for pretreatment of petroleum laden soils may include a crusher, sieve and a mechanical scrubber for breakdown and removal of large soil particles.
- Extraction and separation of soils—For extraction of petroleum hydrocarbons and their separation from the soils equipment includes mixers, settlers, a vacuum belt filter, hydrocyclones, and rotating sieves.
- Post-treatment of soil—Post-treatment of soil includes additional wash-

ings, neutralization if needed, heating for evaporation of organic extracting agents. Equipment may consist of a dewatering screen with washing installation, chemical sprayers, and a heating apparatus.
- Separation of fines from solvent—Fine particles are separated from the extracting agent with sedimentators, hydrocyclones, centrifuges, and flotation units.
- Solvent treatment—Treatment of the extracting agent for reuse involves methods such as carbon adsorption, distillation, etc. and, consequently, the equipment to implement these methods.

Treatment Needs

In order for the chemical extraction to be economically feasible, chemical solvent or surfactant recovery is essential. If an aqueous surfactant is used, then ion exchange, activated carbon absorption, reverse osmosis, microfiltration, stripping, or precipitation are methods which may be considered to separate the hydrocarbons from the solvent. For an organic solvent, evaporation/stripping, distillation, or secondary extraction may be considered. Once the solvent is recovered, it can be recycled back into the process (Smith, 1985).

The extracted soil must be further treated to remove residual solvents so that it can be used as clean backfill. If this is not possible the soil would have to be disposed of based upon the remaining level of solvent and other site-specific conditions.

Disposal Needs

Following extraction and separation of the solvent, the hydrocarbon-rich effluent must then be disposed/treated/recovered at a licensed facility. Possibilities include incineration, product recovery, use as a supplemental fuel, or wastewater treatment.

Monitoring Requirements

Monitoring for cleanup effectiveness should be accomplished by sampling and analyzing soil prior to and after extraction. After the hydrocarbons have been extracted from the soils, the soils may be suitable for "clean" backfill.

Permitting Requirements

Permits/approvals from several government and regulatory agencies may be required. On the local level, the treatment system may require compliance with building and land use ordinances. Water discharged (if any) may require compliance with the National Pollutant Discharge Elimination System (NPDES) standards.

If the recovered hydrocarbons are used as a fuel supplement certain approvals/permits may be needed for the boiler prior to use of these materials. Approvals will also be needed for disposition/backfill of treated soils.

Environmental Feasibility of Chemical Extraction

Exposure Pathways

During excavation, installation, operation, and monitoring of a non-in situ chemical extraction process, possible primary pathways for direct human exposure include the following:

- Inhalation of vapors from the soil or groundwater during excavation, installation, and monitoring of the system and/or inhalation of emissions from the treatment unit during operations and monitoring. When certain compounds combine with the extracting agents, hazardous constituents may be formed. For example, hydrogen cyanide (HCN) is formed as cyanide contacts acids. Safety precautions must be taken to avoid inhalation of potentially harmful gases or skin contact with corrosive liquids (Smith, 1985).
- Particle inhalation and/or ingestion during excavation and system installation.
- Contact of skin with soil or groundwater during excavation and installation. The chemical extraction process itself may involve acids, bases, and detergents which are added to the solvent and are present in the treated soil stream. The primary environmental exposure pathways during excavation, installation, operation, and monitoring for the chemical extraction process include groundwater and surface water. Impacts of the petroleum hydrocarbons and chemical agents used in the process on groundwater and surface water may determine the regulatory cleanup levels if the waters have been affected. During operation and monitoring, contact with the treated soil stream must be avoided.

Secondary pathways include particle inhalation and ingestion during operation and monitoring activities.

Environmental Effectiveness

The effectiveness of the chemical extraction process is highly dependent upon site-specific conditions. As mentioned earlier, soil characteristics, petroleum product composition, solvent characteristics and a variety of other factors all influence the effectiveness of extraction processes. Bench-scale and pilot testing have not completely evaluated the environmental effectiveness of this technology at this time. This technology is still in the development process.

Economic Feasibility of Chemical Extraction*

Capital Costs

Capital costs for this technology are highly variable depending upon site-specific constraints. A custom-designed extraction unit will most likely be required after field feasibility testing has been completed. An estimate for the design and construction of a system to accomplish the remediation for a relatively small spill site is $100,000 to $500,000.

More likely a transportable unit will be leased. Such a unit can be delivered and can process petroleum-bearing soils for $78.43 to $326.80 per cubic meter ($60 to $200 per yd^3) depending on site-specific conditions such as ease of access and excavation. This price includes equipment and labor (Cody, 1986).

Installation Costs

Installation costs are a function of site-specific factors including availability of erection site, distance to utilities, type and size of system to be installed, and other factors that prohibit an accurate estimation of installation cost.

Operation and Maintenance Costs

Routine maintenance costs can be expected to be as much as 5 to 10 percent of the capital costs of $5,000 to $50,000, but will be included in the cost of

*All costs are presented in 1986 U.S. dollars unless otherwise specified.

a leased mobile unit. The cost of the solvent or surfactant will depend on type and usage. Typical surfactant costs are $1.43 to $1.94 per kg ($.65 to $.88/lb). Costs for excavation of soils will be incurred (see the subsection, Excavation).

Qualitative Ranking of Cost

Due to the handling and processing requirements of this technology, the overall cost ranks moderately to relatively high.

EXCAVATION

General Description

Excavation, as referred to in this document, is the process by which the affected soils are removed from the site for disposal. This process, as well as its feasibility, is described in detail in this subsection.

Although excavation and disposal were widely used in the past for removing soils affected by hydrocarbon compounds, today it is generally considered a storage, not a treatment, process and raises issues of future liability for the responsible parties regarding the ultimate disposal of the soils.

Excavation is, however, a major element for most of the non-in situ remedial actions described in this document and is, therefore, discussed here.

Process Description

The excavation process is self-explanatory. The process of landfilling will be described briefly. Sanitary landfills generally operate under the trench method, the area filling method, or some combination of the two (O'Leary et al., 1986).

Material to be disposed of by the trench method is spread and compacted in an excavated trench. Excavated soils are used as the cover material. Cohesive soils such as glacial till or clayey silt are best for use with the trench method. The trench method is generally used when groundwater is low and soils are deep. The method works best on flat or rolling land.

The area-filling method can be applied on most terrain types and is usually utilized when large quantities of material must be disposed of or when excavation below grade is not desirable. The material to be disposed of is spread on the ground surface and covered and compacted in place.

Landfills are typically equipped with impermeable liners or are sited upon some low permeability geologic formation. This is an effort to reduce the risk of groundwater impacts caused by leachate which is generated as the result of natural microbiological action and the infiltration of rainfall. Landfills are designed with runon and runoff controls to minimize leachate generation. Newer landfills are constructed with a leachate collection and treatment system. Closed landfills are capped with some low permeability material to further reduce the infiltration of precipitation.

Technical Feasibility of Excavation

Technical Description

Traditionally, excavation and removal of soils containing hydrocarbons has been an often-used site remediation technique because it allows for immediate and total site cleanup. However, high costs and issues involving the ultimate disposal of excavated soils have encouraged the development of innovative and alternative soil remediation techniques such as those discussed in other sections of this document. Much excavated soil is disposed in various land-fill configurations ranging from municipal sanitary landfills to secure hazardous waste facilities. Because landfilling has traditionally been associated with excavation, this subsection discusses excavation techniques in general with some specific references to landfill disposal practices.

In order to discuss the long-term implications of landfill disposal of excavated materials, the degradation mechanisms within a landfill must be understood.

Waste materials buried in a sanitary landfill are degraded by microorganisms. Figure 4.18 shows three distinct phases of decomposition that occur in a landfill environment (O'Leary et al., 1986). The first stage is aerobic. Degradable solid wastes and oxygen trapped during the landfilling process are metabolized by microorganisms to produce carbon dioxide and water. During this period, temperatures within the landfill will rise and some of the carbon dioxide gas that is generated will dissolve in water to create a weakly acidic environment.

After the oxygen is consumed, a second stage of decomposition occurs as anaerobic facultative organisms, that can live either with or without oxygen, grow in increasing numbers. In the final stages of decomposition, anaerobic organisms utilize hydrogen, carbon dioxide, and organic acids to produce methane gas and other degradation products. Methanogens are very slow growing; therefore, the complete decomposition process occurs over a long period of time.

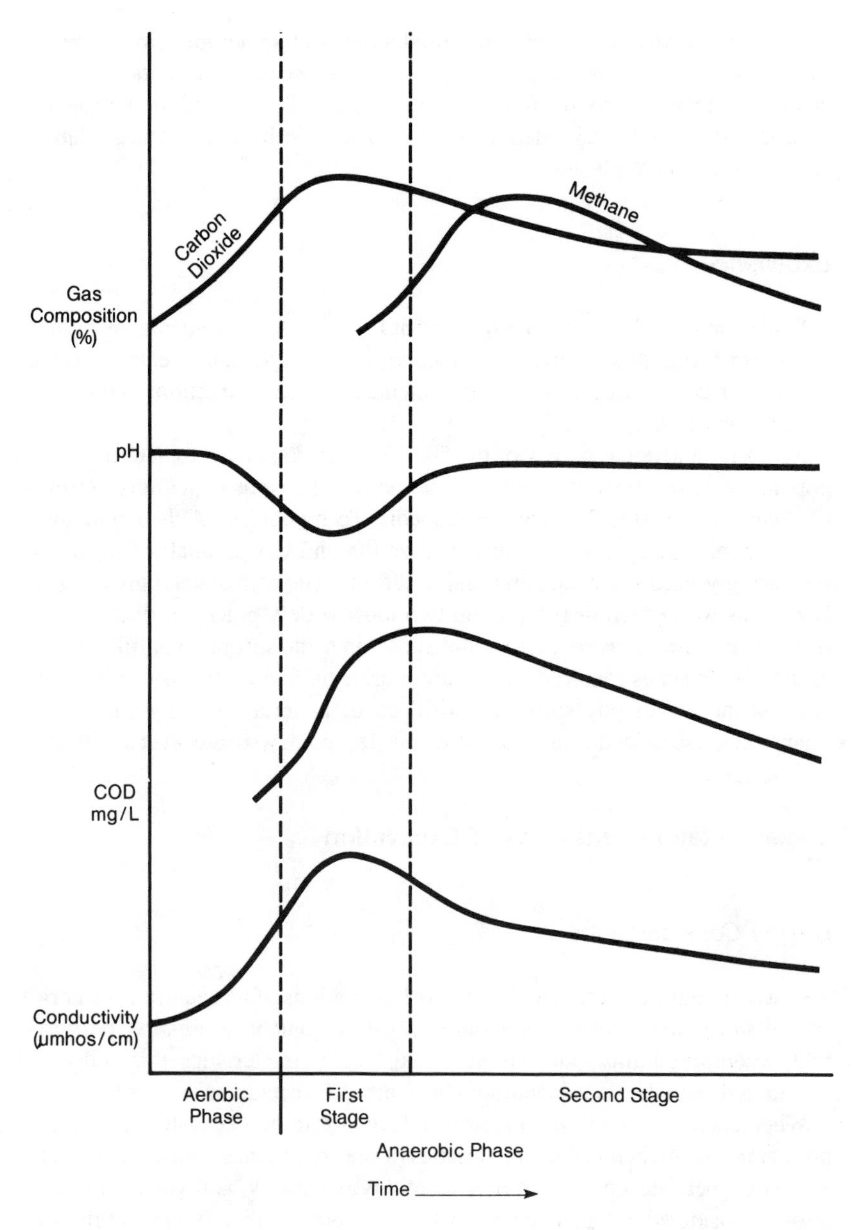

(Source: O'Leary et al., 1986)

Figure 4.18 Three stages of landfill decomposition

Petroleum hydrocarbons are only biodegraded in an aerobic environment; therefore, petroleum products that have been disposed of in a sanitary landfill cannot be expected to be fully degraded. Landfill disposal of petroleum hydrocarbons can be regarded as long-term storage because no degradation is expected to take place.

Experience

Excavation of oily soils is a requirement for many of the remedial technologies described in this report. These technologies include land treatment (land farming or composting), thermal treatment, chemical extraction, and on- or off-site landfilling.

Recently, a national survey of the 50 states was conducted to determine the popularity of various reuse and disposal options for soils containing petroleum hydrocarbons (see Summary of Remedial Technologies). Although innovative technologies such as in situ volatilization and biodegradation are being increasingly used, land farming and landfilling (in both state-approved and hazardous waste facilities) were the two most widely practiced disposal options. Thirty-six states reported soil disposal in some sort of a landfill configuration. Five states required soil aeration prior to landfill disposal. Six states allowed the use of oily soils as landfill cover material. Twenty-nine states reported the use of landfarming as a remedial technology (Kostecki et al., 1985).

Implementation Feasibility of Excavation

Design Considerations

Conventional excavation and earthwork operations often require consideration of soil properties such as texture, moisture content, grain size, permeability, compressibility, and strength. Surface characteristics that must be considered include slope conditions and natural vegetation.

When soils containing petroleum products are to be excavated, some important physical/chemical characteristics of the compounds contained in those soils (i.e., persistence, ignitability, reactivity, solubility, and volatility) must also be considered. Climatic conditions such as precipitation, temperature, and wind speed may affect the decision as to how and when to excavate. Subsurface conditions including the depth to groundwater, subsurface stratigraphy, and aquifer permeability may also be important concerns.

On a national level the U.S. Environmental Protection Agency has developed siting and operational criteria in relation to landfill disposal of municipal solid wastes. Landfill designers must consider floodplains, the preservation of endangered species, the prevention of surface water and groundwater pollution, minimization of disease vectors and air pollution, and the promotion of safety (EPA, 1979).

Equipment Requirements

A large variety of equipment is available for use in activities involving the excavation, removal, and loading of soils. The equipment that is commonly used falls into three general categories:

- Backhoes.
- Cranes and attachments.
- Dozers and loaders.

The backhoe is probably the most commonly used piece of equipment at soil excavation sites. Its greatest advantage is that it can work in tight spaces, causing minimal disturbance to utility lines, buried pipes, and other workers. Conventional backhoes can work to a maximum depth of 7.62 meters (25 feet). Specially designed units are available which can perform work at greater depths. Detailed descriptions of equipment types and capabilities can be found in engineering handbooks, such as *The Excavation Handbook* by H. K. Church (1981).

Other factors affecting the choice of equipment include (EPA, 1985a):

- Equipment efficiency under site-specific conditions.
- Equipment dispatching and setup time.
- Contractor performance record with equipment.
- Equipment idle time.
- Equipment versatility.
- Equipment modifications to increase efficiency and safety.
- Equipment capability for remote handling chores.

Equipment utilized in sanitary landfill operations perform three basic functions:

- Handling of the waste materials.
- Handling of the cover materials.
- Performance of support functions (Brunner and Keller, 1972).

Table 4.14 summarizes the performance characteristics of such equipment (Brunner and Keller, 1972).

Table 4.14 Performance Characteristics of Landfill Equipment

Equipment	Spreading	Compacting	Excavating	Spreading	Compacting	Hauling
Crawler dozer	E	G	E	E	G	NA
Crawler loader	G	G	E	G	G	NA
Rubber-tired dozer	E	G	F	E	G	NA
Rubber-tired loader	G	G	F	G	G	NA
Landfill compactor	E	E	P	E	E	NA
Scraper	NA	NA	G	E	NA	E
Dragline	NA	NA	E	F	NA	NA

Source: Brunner and Keller, 1972

Basis of Evaluation:

1. Easily workable soil
2. Cover material haul distance greater than 305 meters (1,000 ft)

Rating Key:

E—Excellent
G—Good
F—Fair
P—Poor
NA—Not applicable

Treatment Needs

Following excavation, soils can be treated by one of several techniques discussed in Section 4 of this report. Details regarding treatment needs pertaining to these remedial technologies are found in the appropriate subsections.

Disposal Needs

Excavation and removal of soils containing hydrocarbons provides the opportunity for rapid and complete site cleanup. However, it raises the issue of how to ultimately dispose of the excavated soils. Excavated soils can be treated by one of the techniques discussed in this report or they can be disposed of in a landfill.

Monitoring Requirements

Excavation of oily soils should be performed with strict health and safety protocols. Air monitoring should be conducted to determine whether vapor concentrations in the area are high enough to cause an explosive or health hazard.

The lower explosive limit (LEL) or the lower flammable limit (LFL) of a substance is the minimum concentration of a gas or vapor in air below which the substance will not burn when exposed to a source of ignition. Below this concentration, the mixture is too ''lean'' to ignite. The upper explosive limit (UEL) or the upper flammable limit (UFL) of a substance is the maximum concentration of a gas or vapor in the air above which the substance will not burn when exposed to a source of ignition. Above this concentration, the mixture is too ''rich'' to burn or explode (NIOSH et al., 1985). The flashpoint of a substance is the minimum temperature at which it gives off sufficient vapor to form an ignitable mixture. Table 4.15 presents flashpoints and flammable concentration ranges for typical fuel mixtures.

Further air monitoring may be desirable for the purpose of determining the concentrations of specific substances in the air. Portable photoionization meters, flame ionization detectors, infrared analyzers, and detector kits are available for this purpose. In addition, soil sampling and analysis may be required in order to assess to what extent soils should be excavated.

Landfill operators are typically required to perform long-term monitoring and provide closure plans for their facility. Monitoring requirements may include groundwater monitoring and reporting, maintenance and monitoring of waste containment systems, and the presence of security personnel. RCRA-approved facilities must be monitored for 30 years following closure.

Table 4.15 Properties of Some Flammable Fuel Mixtures

	Flash Point Closed Cup °C (°F)	Flammable Limits in Air (mg/L)	
		Lower	Upper
Gas Oil	65.6 (150)	0.5	5.0
Gasoline (general)	7.2 (45)	1.4	7.6
Kerosene	37.8–72.2(100–162)	0.7	5.0
No. 2 Fuel Oil	57.8 (136)	1.3	6.0
No. 6 Fuel Oil	65.6 (150)	1.0	5.0

Source: NIOSH et al.; NFPA, 1985

Permitting Requirements

Landfills are highly regulated because of the large number of potential impacts they may have on human health and the environment. Sanitary landfills used for the disposal of household wastes are specifically exempt from Federal Resource Conservation and Recovery Act regulations. However, the states are passing increasingly stringent landfill regulations in an effort to protect health and environmental safety and to promote alternative disposal options.

Secure hazardous waste facilities must comply with RCRA landfill standards under 40 CFR Part 265. The RCRA regulations provide guidance in regard to location, design, construction, operations, and maintenance of hazardous waste facilities.

Environmental Feasibility of Excavation

Exposure Pathways

Potential direct human exposure pathways during the excavation and transportation of petroleum-bearing soils include the following:

- Vapor inhalation.
- Dust inhalation.
- Particle ingestion.
- Skin contact.

Areas of particular concern are those that frequently experience inversions or that have limited dispersion potential because temperature, wind velocity, precipitation, and topography are all factors that affect the mitigation of dust or gas releases (EPA, 1985b).

Methods and routes of transportation of the soil material affect the potential for human exposure to particulate or gas releases that can be attributed to accidental spills or continuous leaks from vehicles used in the transport of the soils. The risk of exposure via this pathway can be minimized by covering the soil during transport, routing the transport through less populated areas, and by adhering to accident and spill prevention programs and procedures (EPA, 1985b).

Primary environmental exposure pathways include possible impacts to groundwater and surface water. Both can be affected by incomplete excavation practices.

When the ultimate disposal option is landfilling of the soil, there are several potential pathways for exposure. The first pathway is through the surface water. The potential of human exposure is increased when the landfill operation is located near a surface water body or upstream from a drinking water source (EPA, 1985b). If the water body has a high flow rate or a large dilution capacity, the potential for detrimental human exposure can be lessened.

Another potential pathway for human exposure is from subsurface gas releases (EPA, 1985b). Barriers or conduits may exist in the area of the landfill. These features can either serve to decrease or increase the movement of any potentially harmful gases that have developed from the disposed materials.

Lastly, there is a significant potential for human exposure via the groundwater pathway. When the landfill is sited in an area of high subsurface flow rates, high water elevations, high infiltration, high porosity, or the location is upgradient from a potable water source well, the potential magnitude of human exposure is increased (EPA, 1985b). Proper containment and monitoring measures can mitigate potentially detrimental releases to the groundwater.

Environmental Effectiveness

Excavation is often the first step in many of the options available for treatment of soils containing petroleum products. The environmental effectiveness of excavation when followed by land treatment, thermal treatment, solidification, chemical extraction, and asphalt incorporation practices is discussed in the appropriate subject categories in this section. Excavation of the material can be extremely effective in terms of site cleanup because it can be confirmed by field sampling and laboratory analyses that all the petroleum-laden soils have been removed. In the anaerobic environment of the landfill, no significant degradation of the petroleum constituents in the soil occurs, so, although the soil has been removed and disposed of, it has not been effectively treated.

The positive considerations of excavation are that a relatively short time period is required to complete the operation and that complete cleanup is possible. The more negative aspects are the worker/operator safety considerations that are necessary, the short-term impacts of the operation (mainly dust and odor generation and increased runoff), and the relatively high costs associated with the excavation, transportation, and ultimate disposal of the soil (EPA, 1985a).

Disposal Needs

Landfill disposal raises the long-term liability issue. Since hydrocarbons are not fully degraded or destroyed in a landfill environment, the disposer could be held accountable for environmental problems that may occur far into the future. By disposing of material in a municipal or privately operated landfill, the generator gives up control of the material and must accept on good faith the landfill operator's promise to maintain reasonable and reliable operating practices.

Economic Feasibility of Excavation*

Capital Costs

The cost of the actual excavation efforts will depend on excavation depth, site surface characteristics, properties of the petroleum constituents present (such as explosivity) and quantity costs associated with the purchase of excavation equipment (such as backhoes, dozers, and dump trucks) and can range from $25,000 to $100,000 (Environmental Law Institute, 1984). Transportation costs are affected by the distance to the disposal facility, the accessibility of the site, the form of the material to be disposed (i.e., liquid vs. solid), and the amount of material to be moved (Environmental Law Institute, 1984). Disposal costs are set by the tipping fees in effect at the disposal landfill site. It has been observed that for materials containing PCBs, disposal costs will often depend on PCB concentration and the form of the material (liquid vs. solid) (Environmental Law Institute, 1984). All the above costs are also subject to other factors such as community and interstate relations and inflation and regulatory effects. These factors are often difficult to quantify (Environmental Law Institute, 1984).

*All costs are presented in 1986 U.S. dollars unless otherwise specified.

Installation Costs

The excavation and landfill option for disposal of petroleum-bearing soils is a short-term operation, and, as such, does not require any installation costs.

Operation and Maintenance Costs

Estimated costs for leased equipment are \$3.93 per cubic meter (\$3/yd^3) for scrapers, \$2.61 per cubic meter (\$2/yd^3) for backhoes, and \$1.31 per cubic meter (\$1/yd^3) for front-end loaders.

Qualitative Ranking of Cost

Costs associated with landfilling petroleum-laden soils are frequently high because they must include the excavation and transportation costs inherent in the overall operation (Rishel et al., 1982). With increasingly stringent regulatory requirements for transportation and disposal of waste materials, the costs of the overall operation are likely to increase in the future.

SECTION 5

Emerging Technologies

Several new remedial technologies are emerging and show potential for efficiently and cost effectively treating petroleum-laden soils and groundwater. These technologies are in several stages of development ranging from research and development (R&D) to pilot testing and commercial availability. Table 5.1 describes the emerging technologies for both in situ and non-in situ applications, their availability, and their sponsor or manufacturer.

Table 5.1 Emerging Remedial Action Technologies

Remedial Technology	Emerging Technology	Description	Availability/Manufacturer
IN SITU TECHNOLOGIES			
Isolation/ Contaminant	Bottom Sealing	Grout is injected through a series of holes to form a horizontal or curved barrier which forms the bottom of a containment cell.	Not commercially available
Biodegradation	Genetic Engineering	Microbial species are genetically engineered to decompose target compounds.	R/D stage
	Anaerobic Degradation	Anerobic microbial species and conditions are developed to enhance utilization of target compounds.	Pilot stage
Thermal Treatment	Radio-frequency Destruction	Radio frequency electrodes placed along the ground-surface heat	Research at Illinois Institute of Technology

Table 5.1, continued

Remedial Technology	Emerging Technology	Description	Availability/Manufacturer
IN SITU TECHNOLOGIES			
Thermal Treatment (continued)		the subsurface and volatilize and/or destroy organics.	
Leaching	Supercritical Leaching	Supercritical fluids are used to leach organics from soil.	Pilot scale unit available: Critical/Fluid Systems, Inc. Cambridge, Massachusetts 617-492-1631
NON-IN SITU TECHNOLOGIES			
Pump and Treatment for Groundwater	Genetic Engineering	Microbial species are genetically engineered to utilize target compounds.	R/D stage
	Combined Technologies	Microbial species are used to bioregenerate spent carbon adsorption beds.	Research stage
	Anaerobic Degradation	Anaerobic microbial species and conditions are developed to enhance utilization of target compounds.	Pilot stage
Thermal Treatment	Electric Reactors	Electrically heated fluid wall reactors are used to pyrolize organics from soils.	Commercial units under construction: Thagard Research Corporation,Costa Mesa, California 714-556-4470
	Infrared Incinerators	Ceramic-lined chamber is heated with silicon carbide heating elements to provide infrared energy.	Several operational units: Shirco Infrared Systems Dallas, Texas 214-630-7511
	Molten Glass	Organics are destroyed, volatilized, and immobilized by the cooling molten glass.	Commercially available: Penberthy Electroment International Seattle, Washington 206-762-4244

Table 5.1, continued

Remedial Technology	Emerging Technology	Description	Availability/Manufacturer
Thermal Treatment (continued)	Molten Salt	Organics are volatilized or destroyed by contact with molten salt.	Pilot-scale units available: Rockwell International Ontario, California 213-700-8200
	Supercritical Water Oxidation	Water at pressure above 224 kg/cm² (3,200 psi) and temperatures above 500°C (932°F) are used to remove organics from soil.	Construction of commercial unit in process: Vertox Corporation Dallas, Texas
	Plasma System	A plasmatic arc generating temperatures approaching 10,000°C is used to destroy organics.	Pilot testing on liquids in progress: Pyrolysis System, Inc. El Segundo, California 416-735-2401
Land Treatment	Soil Shredding	Soils are shredded and aerated to volatilize organic compounds.	Commercially available: SMC Martin, Inc. Valley Forge Pennsylvania 215-299-6000
Chemical Extraction	Supercritical Extraction	Supercritical carbon dioxide is used to extract organics from soils.	Laboratory testing stage: Critical Fluid Systems, Inc. Cambridge, Massachusetts 617-492-1631
	Ultraviolet Photolysis	Extracted organics are destroyed or transformed by UV irradiation.	Laboratory scale: SYNTEX
Excavation	Soil Freezing	Soil is frozen to facilitate excavation of adsorbed organics.	Available through construction contractors

References

SECTION 2—SUMMARY OF REMEDIAL TECHNOLOGIES

M. Bonazountas, J. Wagner, and M. Alsterburg. "Potential Fate of Buried Halogenated Solvents via SESOIL." Published by A.D. Little, Inc., January 1983.

M. Bonazountas and J. Wagner. "SESOIL: A Seasonal Soil Compartment Model." A.D. Little, Inc. for the United States Environmental Protection Agency, May 1984.

M. Bonazountas, J. Wagner, and B. Goodwin. "Evaluation of Seasonal Soil/Groundwater Pollutant Pathways via SESOIL." Published by A.D. Little, Inc., July 1981.

E.J. Fleischer, P.R. Noss, P.T. Kostecki, and E.J. Calabrese. "Evaluating the Subsurface Fate of Organic Chemicals of Concern Using the SESOIL Environmental Fate Model." In *Proceedings of the Third Eastern Regional Groundwater Conference,* 29–31 July 1986, Springfield, Massachusetts, National Water Well Association.

P.T. Kostecki, E.J. Calabrese, and E. Garnick. "Regulatory Policies for Petroleum Contaminated Soils: How States Have Traditionally Dealt with the Problem." In *Proceedings of the Conference on the Environmental and Public Health Effects of Soils Contaminated with Petroleum Products*, University of Massachusetts, Amherst, Massachusetts, 30–31 October 1985, (in press).

D.B. Watson and S.M. Brown. "Testing and Evaluation of the SESOIL Model." Anderson-Nichols, Inc. for the U.S. Environmental Protection Agency, September 1984.

SECTION 3—IN SITU TECHNOLOGIES

In Situ Volatilization

A. Baehr, and M.Y. Corapcioglu. "A Predictive Model for Pollution from Gasoline in Soils and Groundwater." In *Proceedings of Petroleum Hydrocarbons and Organic Chemicals in Groundwater*, 5–7 November 1984, Worthington, Ohio, National Water Well Association.

Brock, T.D., Smith, P.W., Madigan, M.T., *Biology of Microorganisms,* 4th ed., Prentice Hall, Inc., Englewood Cliffs, NJ, 1984, p. 847.

M.F. Coia, M.H. Corbin, G. Anastos, and D. Koltuniak. "Soil Decontamination Through In Situ Air Stripping of Volatile Organics—A Pilot Demonstration." Presented at Petroleum Hydrocarbons and Organic Chemicals in Ground Water, 13–15 November 1985, NWWA, Houston, Texas.

W.A. Jury, "*Volatilization From Soil*," and S.C. Hern and S.M. Melancon, "Guidelines for Field Testing Soil Fate and Transport Models-Final Report, Appendix B." EPA 600/4-86-020. April 1986.

G.E. Hoag and B. Cliff. "The Use of the Soil Venting Technique for the Remediation of Petroleum Contaminated Soils." Civil Engineering Department, University of Connecticut, Storrs, Connecticut, 1985.

M.J. O'Conner, J.G. Agar, and R.D. King. "Practical Experience in the Management of Hydrocarbon Vapours in the Subsurface." In *Proceedings of the NWWA/API Conference on Petroleum Hydrocarbons and Organic Chemicals in Ground Water—Prevention, Detection and Restoration,* 5–7 November 1984, Houston, Texas.

In Situ Biodegradation

American Petroleum Institute. *Cost Model for Selected Technologies for Removal of Gasoline Components from Groundwater*. Doc. No. 4442, February 1986.

R.M. Atlas "Microbial Degradation of Petroleum Hydrocarbons: An Environmental Perspective," *Microbial Review*, Vol. 45, 1981, pp. 180–209.

R. Bartha and R.M. Atlas. "The Microbiology of Aquatic Oil Spills." *Advances in Applied Microbiology*, Vol. 22, 1977, pp. 225–266.

I. Bossert and R. Bartha. "The Fate of Petroleum in Soil Ecosystems." *Petroleum Microbiology*, R.M. Atlas, ed., Macmillan Co., New York, 1984, pp. 435–473.

R.S. Brown, R.D. Norris, and M.S. Westray. "In Situ Treatment of Groundwater." Presented at HazPro '86: The Professional Certification Symposium and Exposition, 1–4 April, Baltimore, Maryland.

J.T. Dibble and R. Bartha. "Effect of Environmental Parameters on Biodegradation of Oil Sludge." *Applied Environmental Microbiology*, Vol. 37, 1979, pp. 729–739.

P.C. Grady. "Biodegradation: Its Measurement and Microbiological Basis." *Biotechnology and Bioengineering*, Volume 27, 1985, pp. 660–674.

E. Heyse, S.C. James, and R. Wetzel. "In Situ Aerobic Biodegradation of Aquifer Contaminants at Kelly Air Force Base." *Environmental Progress*, August 1986, pp. 207–211.

R.L. Huddleston and L.W. Cresswell. "Environmental and Nutritional Constraints of Microbial Hydrocarbon Utilization in the Soil." In *Proceedings of the 1975 Engineering Foundation Conference: The Role of Microorganisms in the Recovery of Oil.* NSF/RANN, Washington, DC, 1975, pp. 71–72.

A. Jobson, M. McLaughlin, F.D. Cook, and D.W.S. Westlake. "Effect of Amendments on the Microbial Utilization of Oil Applied to Soil." *Applied Microbiology*, Vol. 27, 1974, pp. 166–171.

C.B. Kincannon. *Oily Waste Disposal by the Soil Cultivation Process*. EPA Report No. R2-72-110, 1972.

M.D. Lee and C.H. Ward. "Reclamation of Contaminated Aquifers: Biological Techniques." Presented at the 1984 Hazardous Spills Conference.

M. Lehtomaki and S. Niemela. "Improving Microbial Degradation of Oil in Soil." *Ambio*, Vol. 4, 1975, pp. 126–129.

National Academy of Sciences. *Fate of Petroleum in the Marine Environment*. National Academy of Sciences, Washington, DC, 1984.

R.L. Raymond, J.O. Hudson, and V.W. Jamison. "Oil Degradation in Soil." *Applied Environmental Microbiology*, Vol. 31, 1976, pp. 522–535.

P.H. Roux. "Impact of Site Hydrogeology on In Situ Remediation Strategies." Aquifer Remediation Systems, FMC, undated.

R. Vanloocke, R. DeBorger, J.P. Voets, and W. Verstraete. "Soil and Groundwater Contamination by Oil Spills: Problems and Solutions." *International Journal of Environmental Studies*, Vol. 8, 1975, pp. 99–111.

W. Verstraete, W., R. Vanloocke, R. DeBorger, and A. Verlinde. "Modeling of the Breakdown and the Mobilization of Hydrocarbons in Unsaturated Soil Layers." In *Proceedings of the 3rd International Biodegradation Symposium*. J.S. Sharpley and A.M. Kaplan, eds., Applied Science Publishers, London, 1976, pp. 99–112.

J.T. Wilson and J.F. McNabb. "Biological Transformation of Organic Pollutants in Groundwater." *EOS*, Vol. 64, No. 33, August 1983, pp. 505–507.

J.T. Wilson, L.E. Leach, M. Henson, and J.N. Jones. "In Situ Biorestoration as a Ground Water Remediation Technique." *Ground Water Monitoring Review*, Fall 1986, pp. 56–64.

P.M. Yaniga, and W. Smith. "Aquifer Restoration: In Situ Treatment and Removal of Organic and Inorganic Compounds." American Water Resources Association, August, 1985.

Roy F. Weston, Inc. "Underground Storage Tank Leakage Prevention Detection and Correction." *Report for Petroleum Marketers Association of America*, August, 1986.

In Situ Leaching and Chemical Reaction

E.F. Dul, R.T. Fellman, M.F. Kuo, and J.F. Roetzer. "Land Disposal of Hazardous Waste." In *Proceedings of the Tenth Annual Research Symposium*. Ft. Mitchell, Kentucky, 1984.

E.D. Ellis, J.R. Payne, and G.D. McNabb. "Treatment of Contaminated Soils with Aqueous Surfactants." EPA/600/S2-85/129, 1985.

V.E. Farmer, Jr., "Behavior of Petroleum Contaminants in an Underground Environment." In *Proceedings of the Seminar on Groundwater and Petroleum Hydrocarbons; Protection, Detection, Restoration*. Petroleum Association for Conservation of the Canadian Environment, Toronto, 1983.

N.M. Johnson. "Assessing the Capabilities of Microbial Degradation of Used Lubricating Oil on Soil." Master's Thesis, Graduate College of the University of Illinois at Urbana-Champaign, Urbana, Illinois, 1986.

Texas Research Institute. *Underground Movement of Gasoline on Groundwater and Enhanced Recovery by Surfactants*. Final Report. Submitted to the American Petroleum Institute, Austin, Texas, 1979.

Texas Research Institute. *Test Results of Surfactant Enhanced Gasoline Recovery in a Large-Scale Model Aquifer.* Report submitted to the American Petroleum Institute, Austin, Texas, 1985.

United States Environmental Protection Agency. *Review of In-Place Treatment Techniques for Contaminated Surface Soils; Volume 1: Technical Evaluation*. EPA/540/2-84-003, Washington, DC, 1984a.

United States Environmental Protection Agency. Handbook: *Remedial Action at Waste Disposal Sites*. EPA/625/6-85/006, Washington, DC, 1985.

In Situ Vitrification

R.A. Brouns, J.L. Buelt, and W.F. Bonner. "In Situ Vitrification of Soil." U.S. Patent No. 4,376,598, March 1983.

J.L. Buelt, V.F. FitzPatrick, and C.L. Timmerman. "Electrical Technique for In-Place Stabilization of Contaminated Soils." Chemical Engineering Progress, March 1985, pp. 43–48.

V.F. FitzPatrick, J.L. Buelt, K.H. Oma, and C.L. Timmerman. "In Situ Vitrification: A Potential Remedial Action for Hazardous Waste." Presented at the Hazardous Material Conference, Philadelphia, Pennsylvania, PNL-SA-12316, 5 June 1984.

T. Larsen, and W.A. Lanford. "Hydration of Obsidian." Nature, Vol. 276, No. 9, 1978, pp. 153–156.

J.H. Oma, D.R. Brown, J.L. Buelt, V.F. FitzPatrick, K.A. Hawley, G.B. Mellinger, B.A. Napier, D.J. Silviera, S.L. Stein, and C.L. Timmerman. "In Situ Vitrification of Transuranic Wastes: Systems Evaluation and Applications Assessment." PNL-4800, Pacific Northwest Laboratories, Richland, Washington, 1983.

Pacific Northwest Laboratories. *Application of In Situ Vitrification to PCB-Contaminated*

Soils. Prepared for Electric Power Research Institute, EPRCS-4834, RP1263-24, October 1986.

Timmerman, C.L., Senior Research Engineer, Pacific Northwest Laboratories, Personal Communication, 29 September 1986.

In Situ Passive Remediation

S.T. Brookman, et al. "Literature Survey: Unassisted Natural Mechanisms To Reduce Concentrations Of Soluble Gasoline Components." TRC Environmental Consultants, Inc., 7 August 1985, API No. 4415.

L.W. Canter and R.C. Knox. *Ground Water Pollution Control*. Lewis Publishers, Inc. 1985.

C.T.I. Odu. "Microbiology of Soils Contaminated with Petroleum Hydrocarbons. I. Extent of Contamination and Some Soil and Microbial Properties After Contamination." *Journal of the Institute of Petroleum*, Vol. 58, No. 562, July 1972.

In Situ Isolation/Containment

J. Ayers, et al. "The First EPA Superfund Cutoff Wall, Design and Specifications." Presented at the Third National Symposium on Aquifer Restoration and Groundwater for Monitoring, May 1983.

F.J. Barnes, J.C. Rodgers, and G. Trujillo. "The Role of Water Balance In the Long-Term Stability of Hazardous Waste Site Cover Treatments." Los Alamos National Laboratory, Los Alamos, New Mexico, 1986.

J. Ehrenfelder and J. Bass. "Handbook for Evaluating Remedial Action Plans." EPA-600/2-83-076. A.D. Little Inc. Prepared for U.S. EPA, Municipal Environmental Research Laboratory, Cincinnati, Ohio, 1983.

H. Glick. "Coal Ash Used For Landfill Cover." *World Wastes*, December 1984.

J.D. Guertin and W.H. McTigue. "Groundwater Control Systems for Urban Tunneling." FHWA/RD-81/073, Vol. 1, Goldberg-Zoino and Associates, Inc. Prepared for U.S. Department of Transportation, Federal Highway Administration, Washington, DC, 1982.

Hayward Baker Company, EarthTech Research Corporation, and ENSCO, Inc. *Chemical Soil Grouting. Improved Design and Control (Draft)*. Federal Highway Administration, Fairbank Research Station, McLean, Virginia, 1980.

F.L. McGarry and B.L. Lamarre. "Groundwater Reclamation, Gilson Road Hazardous Waste Disposal Site, Nashua, New Hampshire." Report, Roy F. Weston, Inc., 1985.

Charles A. Moore. "Landfill and Surface Impoundment Performance Evaluation Manual." SW-869, Geotechnics, Inc. Prepared for U.S. EPA, Municipal Environmental Research Laboratory, Cincinnati, Ohio, 1980.

Office of Research and Development, U.S. EPA, Municipal Environmental Research Laboratory, Cincinnati, Ohio. *Slurry Trench Construction for Pollution Migration Control*. EPA-540/2-84-001, February 1984.

Office of Research and Development. U.S. EPA Hazardous Waste Engineering Research Laboratory, Cincinnati, Ohio. Handbook: *Remedial Action At Waste Disposal Sites (Revised)*. EPA-625/6-85/006, October 1985.

N. Peters, et al. "Applicability of the HELP Model In Multilayer Cover Design: A Field Verification and Modeling Assessment." Agricultural Engineering Department, University of Kentucky, Lexington, Kentucky. Undated.

L.T. Schaper and G.A. Neeley. "Synthetic Foam Cover for Sanitary Landfills: An Examination and Update." *Public Works*, May 1985.

S. Sommerer and J.F. Kitchens. "Engineering and Development Support of General Decon Technology for the DARCOM Installation Restoration Program." *Task 1 Literature Review on Groundwater Containment and Diversion Barriers (Draft)*. Atlantic Research Corporation. Prepared for: U.S. Army Hazardous Materials Agency, Aberdeen Proving Ground, Maryland, 1980.

SECTION 4—NON-IN SITU TECHNOLOGIES

Land Treatment Technology

American Petroleum Institute. *Landfarming: An Effective and Safe Way to Treat/Dispose of Oily Refinery Wastes*. Solid Wastes Management Committee, March 1980.

American Petroleum Institute. *Land Treatability of Appendix VIII Constituents Present in Petroleum Industry Wastes*. Document B974220, May 1984.

Ronald M. Atlas. "Microbial Degradation of Petroleum Hydrocarbons: An Environmental Perspective." *Microbiological Reviews*, 45(1), March 1981.

K.W. Brown, L.E. Deuel, and J.C. Thomas. "Land Treatability of Refinery and Petrochemical Sludges." Municipal Environmental Research Lab, Cincinnati, Ohio, August 1983.

J.T. Dibble and R. Bartha. "Effect of Environmental Parameters on the Biodegradation of Oil Sludge." *Applied and Environmental Microbiology*, April 1979.

J.T. Dibble and R. Bartha. "Rehabilitation of Oil Inundated Agricultural Land: A Case History." *Soil Science*, 128(1), July 1979.

H.E. Knowlton and J.E. Rucker. "Landfarming Shows Promise for Refinery Waste Disposal." *The Oil and Gas Journal*, 14 May 1979.

D.J. Norris. "Landspreading of Oily and Biological Sludges in Canada." In *Proceedings of the 35th Industrial Waste Conference*. Purdue University, 14–15 May 1980.

K. Perlin and E. F. Gilardi. "Review of European Composting Systems." Presented at the National Conference of Design of Municipal Sludge Compost Facilities, Chicago, 29–31 August 1978.

Tan, L. Phung, Barker, D. Ross, and D. Bauer. "Land Cultivation of Industrial Wastes and Municipal Solid Wastes: State of the Art Study." EPA-60012-78-140a, August 1978.

U.S. EPA. *Composting of Municipal Wastewater Sludges*. EPA/625/4-85/014, August 1985.

Thermal Treatment

Samuel Hankin. "EPA's Incinerator Passes First Test." *World Wastes*, August 1985.

P.T. Kostecki, E.J. Calabrese, and E. Garnick. "Regulatory Policies for Petroleum Contaminated Soils: How States Have Traditionally Dealt With the Problem." University of Massachusetts, Amherst, Massachusetts, 1985.

John W. Noland, Nancy P. McDevitt, and Donna L. Koltuniak. "Low Temperature Thermal Stripping of Volatile Compounds—A Field Demonstration Project." Presented at the 18th Mid-Atlantic Industrial Waste Conference, Blacksburg, Virginia, June 1986.

William S. Rickman, Nadine D. Holder, and Darrell T. Young. "Circulating Bed Incineration of Hazardous Wastes." Chemical Engineering Progress, March 1985.

U.S. Environmental Protection Agency. Handbook: *Remedial Action of Waste Disposal Sites (Revised)*. EPA/625/6-85-006, 1985.

D.L. Vrable and D.R. Engler. "Transportable Circulating Bed Combustor (CBC) for the Incineration of Hazardous Waste." Presented at HazPro '86, Baltimore, Maryland, April 1–3, 1986.

Asphalt Incorporation

Asphalt Institute. *Principles of Construction of Hot-Mix Asphalt Pavements*. Manual Series No. 22, January 1983.

Stanley Bemben. Professor, Civil Engineering Department, University of Massachusetts, Amherst, Massachusetts, Personal Communication, 12 September 1985.

Bob Joubert. N.E. Regional Engineer, Asphalt Institute, Personal Communication, 15 August 1985.

P.T. Kostecki, E.J. Calabrese, and E. Garnick. "Regulatory Policies for Petroleum Contaminated Soils: How States Have Traditionally Dealt With the Problem." In *Proceedings of the Conference on the Environmental and Public Health Effects of Petroleum Contaminated Soils*. University of Massachusetts, Amherst, Massachusetts, 30–31 October 1985, (in press).

Chuck Pagan, Director of Research, National Asphalt Paving Association, Personal Communication, 20 August 1985.

Solidification/Stabilization

American Society for Testing and Materials. Annual Book of ASTM Standards, Philadelphia, Pennsylvania, 1985.

M.J. Cullinane. "Field Scale Solidification/Stabilization of Hazardous Waste." In *Proceedings of the 1985 ASCE Specialty Conference on Environmental Engineering*. Northeastern University, Boston, Massachusetts, 1–5 July, 1985.

M.J. Cullinane and L.W. Jones. "Technical Handbook for the Stabilization/Solidification of Hazardous Waste." Prepared for the U.S. EPA Hazardous Waste Engineering Laboratory, Cincinnati, Ohio, 1985.

Electric Power Research Institute. *Coal Combustion By-Products Utilization Manual*. EPRI CS-3122. July 1983.

Environmental Protection Agency. *Guide to the Disposal of Chemically Stabilized and Solidified Waste*. SW-872. September, 1980.

B. Mather. "Pozzolan: Reviews in Engineering Geology." D.J. Varnes and G. Kiersch, eds. The Geological Society of America, 1969.

P.K. Mehta. "Pozzolanic and Cementitious By-Products as Mineral Admixtures for Concrete—A Critical Review." in *Fly Ash, Silica Fume, Slag, and Other Mineral By-Products in Concrete*. Vol. 1. V.M. Malhotra, ed. American Concrete Institute, 1983.

D.S. Morgan, J.I. Novoa, and A.H. Halff. "Oil Sludge Solidification Using Cement Kiln Dust." *ASCE Journal of Environmental Engineering*. Vol. 110, No. 5, October 1984, pp. 935–948.

D.T. Musser,and R.L. Smith. "Case Study: In Situ Solidification/Fixation of Oil Field Production Fluids—A Novel Approach." In *Proceedings of the 39th Industrial Waste Conference*. Purdue University, 8–10 May 1984.

C.L. Smith. IU Conversion Systems, Inc., Personal Communication, August 1986.

C.L. Smith and T. Longosky. "Operating Experiences in Hazardous Wastes Disposal by Pozzolanic Cementation." In *Hazardous and Toxic Wastes: Technology, Management, and Health Effects*. S.K. Majumdar and E.W. Miller, eds. Pennsylvania Academy of Science, 1984.

C.L. Smith and K.E. Zenobia. "Pozzolanic Microencapsulation for Environmental Quality Assurance." In *Proceedings of the 37th Industrial Waste Conference*. Purdue University, 11–13 May 1982.

M.K. Sonksen and J.A. Lease. "Evaluation of Cement Dust Stabilization of Polychlorinated Biphenyl Contaminated Sludges." In *Proceedings of the 37th Industrial Waste Conference*. Purdue University, 11–13 May 1982.

M.E. Tittlebaum, F.K. Cartledge, and S. Engels. "State of the Art on Stabilization of Hazardous Organic Liquid Wastes and Sludges. "*CRC Critical Reviews in Environmental Control*. Vol. 15, No. 2, 1985, pp. 179–211.

Transportation Research Board. *Lime-Fly Ash: Stabilized Bases and Subbases*. TRB-NCHRP Synthesis Report 37, 1976.

Groundwater Extraction and Treatment

Fram Industrial. *Coalescing Plate Separators for Environemntal Cleanup and Resource Recovery*. 8-82/20M/TLFA. Oil Recovery Systems, Inc. *Scavenger "Light Oil" Separator Cartridge*, Company Literature, undated.

D.M. Giusti et al. "Activated Carbon Adsorption of Petrochemicals." *TWPFC*, Vol. 46, No. 5, May 1974.

D.W. Hall. "Carbon Adsorption As An Interim Remedial Measure At Private Water Wells." In *Proceedings of the Petroleum Hydrocarbons and Organic Chemicals in Groundwater—Prevention, Detection and Restoration—A Conference and Exposition*, Houston, Texas, 13–15 November 1985.

D.W. Hand et al. "Design and Evaluation of an Air Stripping Tower for Removing VOCs From Groundwater." *Journal of the American Water Works Association*, September 1986.

A.R. Jacobson. "Granular Activated Carbon Solves Water Emergency." *Public Works*, January 1981.

M.C. Kavanaugh and R.R. Trussell. "Design of Aeration Towers to Strip Volatile Organic Contaminants from Drinking Water." *Journal of the American Water Works Association*, December 1980.

R.P. O'Brien and J.L. Fisher. "There is an Answer to Groundwater Contamination." *Water/Engineering and Management*, May 1983.

Oil Recovery Systems, Inc. "Innovative Equipment and Removal Techniques for Removal of Pollutants from Groundwater and Soils." 1986.

Process Design Manual. "Wastewater Treatment Facilities for Sewered Small Communities." EPA-625/1-77-009. October 1977.

Chemical Extraction

B.A. Beaudet, J.W. Vinzant, L., J. Bilello, and J.D. Crane. "Development of Optimum Treatment Systems for Industrial Wastewater Lagoons, Task 3, Volume I." DAAK11-81-C-0076. Environmental Science and Engineering, Inc., Gainesville, Florida, June 1983.

Lawrence J. Bove et al. "Removal of Contaminants From Soil, Phase I: Identification and Evaluation of Technologies." Prepared for U.S. Army Toxic and Hazardous Materials Agency, Aberdeen Proving Ground, Maryland. December 1983.

T. Cody. Marketing Critical Fluid Systems, Personal Communication, 14 October 1986.

K.R. Huibregste et al. "Development of a Mobile System for Extracting Spilled Hazardous Materials From Soils." 1980 Hazardous Materials Conference.

James Nash and Richard P. Traver. "Field Evaluation of In Situ Washing of Contaminated Soil with Water/Surfactants," undated.

R. Scholz and J. Milanowski. "Mobile System for Extracting Spilled Hazardous Materials From Excavated Soils." Project Summary, United States Environmental Protection Agency, EPA-600/S2-83-100, December 1983.

M.A. Smith. *Contaminated Land*. Plenum Press, New York 1985.

Excavation

D.R. Brunner and D.J. Keller. "Sanitary Landfill Design and Operation." U.S. EPA (SW-65ts), 1972.

H.K. Church. *The Excavation Handbook*. McGraw-Hill Book Company, New York, 1981.

U.S. Environmental Law Institute. *Summary Report: Remedial Responses at Hazardous Waste Sites*. EPA-540/2-84-002, March 1984.

U.S. Environmental Protection Agency. "Criterial for Classification of Solid Waste Disposal Facilities and Practices." *Federal Register*, Vol. 44, No. 179, 13 September 1979, pp. 53438–53468.

U.S. Environmental Protection Agency. *Remedial Action at Waste Disposal Sites (Revised)*. EPA/625/6-85/006. October 1985a.

Environmental Protection Agency. *Permit Applicants Guidance Manual for Exposure Information Requirements under RCRA Section 3019*. Office of Solid Waste, 3 July 1985b.

E.J. Kostecki, Calabrese, and E. Garnick, "Regulatory Policies for Petroleum-Contaminated Soils: How States have Traditionally Dealt with the Problem." In *Proceedings of the Conference on the Environmental and Public Health Effects of Soils Con-*

taminated with Petroleum Products. University of Massachusetts, Amherst, Massachussetts, 30–31 October 1985 (in press).

National Fire Protection Association. *Fire Protection Guide on Hazardous Materials*. Eighth Edition, Boston, Massachusetts, 1985.

National Institute for Occupational Safety and Health, Occupational Safety and Health Administration, U.S. Coast Guard, and U.S. Environmental Protection Agency. *Occupational Safety and Health Guidance Manual for Hazardous Waste Site Activities*. U.S. Department of Health and Human Services, Washington, DC, October, 1985.

P.R. O'Leary, L. Canter, and W.D. Robinson. "Land Disposal." In *The Solid Waste Handbook*. W.D. Robinson, ed., John Wiley and Sons, New York, 1986.

H.L. Rishel, T.M. Boston, and C.J. Schmidt. "Costs of Remedial Response Actions at Uncontrolled Hazardous Waste Sites." EPA-600/2-82-035, March 1982.

Index

aerobic biodegradation, 105 (*See also* biodegradation)
anaerobic biodegradation, 107, 195–196
asphalt incorporation
  economic feasibility
    capital costs, 142–143
    installation costs, 143
    operation and maintenance costs, 143
    qualitative ranking of cost, 143
  environmental feasibility
    environmental effectiveness, 142
    exposure pathways, 141–142
  general description, 135
  implementation feasibility
    design, 139–140
    disposal, 140–141
    equipment, 140
    monitoring, 141
    permitting, 141
    treatment, 140
  process description, 135–137
  technical feasibility
    description, 137
    experience, 138
ASTM Method A Extraction Procedure, 153

biodegradation (in situ)
  economic feasibility
    capital costs, 51
    installation costs, 51
    operation and maintenance costs, 51
    qualitative ranking of costs, 51
  environmental feasibility
    environmental effectiveness, 49
    exposure pathways, 48–49
  general description, 37–38
  implementation feasibility
    design, 45–46
    disposal, 47
    equipment, 47
    monitoring, 48
    permitting, 48
    treatment, 47
  in soils, 41–43
  process description, 38
  technical feasibility
    description, 41
    experience
      leaking underground storage tank case study, 43–44
      tank leak case study, 44–45
bottom sealing, 195
bulking agent, 103

capping
  constructed cap (asphalt or concrete), 86–87
  low permeability soils, 86
  multilayer cap, 87
  soil/bentonite admixtures, 86
  synthetic membrane caps, 85
carbon adsorption
  vapor phase, 32–33
cation exchange capacity (CEC), 113
chemical extraction
  economic feasibility
    capital costs, 182
    installation costs, 182
    operation and maintenance costs, 182–183
    qualitative ranking of cost, 183
  environmental feasibility
    environmental effectiveness, 182

exposure pathways, 181
general description, 175–176
implementation feasibility
design, 178–179
disposal, 180
equipment, 179–180
monitoring, 180
permitting, 181
treatment, 180
process description, 176
technical feasibility
description, 176
experience
bench-scale case study, 178
pilot-scale case study, 178
Clean Air Act
National Ambient Air Quality Standards, 171
composting (*See also* land treatment)
in-vessel, 103
process considerations, 113–114
static pile, 103
windrow, 103
Comprehensive Environmental Response Compensation and Liability Act (CERCLA), 131
computer simulation
Seasonal Soil Compartment Model (SESOIL), 8–11
containment (*See* isolation/containment)
cutoff walls, 89

degradation
chemical, 105
microbial, 105–107 (*See also* biodegradation, aerobic biodegradation)
photochemical, 105
delivery methods
forced, 55
gravity, 55
disk tilling, 114

Edison Electric Institute (EEI), 2
Electric Power Research Institute (EPRI)
evaluation of underground storage tank technology, 1
in situ vitrification, 67
electric reactors, 196
enhanced oil recovery (EOR) techniques, 53
excavation
handbook, 187
economic feasibility
capital costs, 192
installation costs, 193
operation and maintenance costs, 193
qualitative ranking of cost, 193
environmental feasibility
disposal, 192
environmental effectiveness, 191–192
exposure pathways, 190–191
general description, 183
implementation feasibility
design, 186–187
disposal, 189
equipment, 187
monitoring, 189
permitting, 190
treatment, 189
process description, 183–184
technical feasibility
description, 184–186
experience, 186
exposure pathways
direct human, 11
environmental, 11
primary, 11
secondary, 11
extraction (*See also* chemical extraction)
ASTM Method A Extraction Procedure, 153
EPA Extraction Procedure, 153
supercritical, 197

fan
forced draft (FD), 121
induced draft (ID), 121
flaring, 32–33
fluidized bed incinerator, 121

genetic engineering, 195–196
groundwater extraction and treatment
economic feasibility
capital costs, 173
installation costs, 173
operation and maintenance costs, 175
qualitative ranking of cost, 175
environmental feasibility
environmental effectiveness, 172

exposure pathways, 172
general description, 157
implementation feasibility
design, 168
disposal, 170–171
equipment, 168–169
monitoring, 171
permitting, 171
treatment, 170
process description, 157–161
technical feasibility
description, 161–165
experience
air stripping case study, 166–167
carbon adsorption case study, 166
grout curtain technique, 89–92

Henry's Law Constant
indicator of stripability, 18
measure of volatilization, 23
hydrocarbons
biodegradation in soils, 41–43
composition, 8, 18
degradation, 104–107
environmental fate, 7–11
transport mechanisms, 20
volatilization determinants, 18–20
hydrogen peroxide
injection in biodegradation system, 38, 44, 46
hydrogeology
importance in biodegradation system design, 45–46
importance in groundwater extraction and treatment design, 161
Hydrologic Evaluation of Landfill Performance (HELP), 97

in situ technologies
biodegradation (*See* biodegradation)
future, 195–196
isolation/containment (*See* isolation/containment)
leaching and chemical reaction (*See* leaching and chemical reaction)
passive remediation (*See* passive remediation)
vitrification (*See* vitrification)
volatilization (*See* volatilization)

isolation/containment (in situ)
economic feasibility
installation costs, 98
maintenance and monitoring costs, 98
qualitative ranking of cost, 99
environmental feasibility
environmental effectiveness, 97
exposure pathways, 97
general description, 84
implementation feasibility
design, 95
disposal, 96
equipment, 96
monitoring, 96
permitting, 96–97
treatment, 96
process description
capping, 85–87 (*See* capping)
groundwater containment, 87–92
grout curtain technique, 89–92
sheet piling, 92
slurry walls/cutoff walls, 89
technical feasibility
description, 92–93
experience
slurry wall and capping case study, 93–95

land treatment
economic feasibility
capital costs, 117
installation costs, 117
operation and maintenance costs, 118
qualitative ranking of cost, 118
environmental feasibility
environmental effectiveness, 116–117
exposure pathways, 116
general description, 101
implementation feasibility
design, 112–114
disposal, 115
equipment, 114
monitoring, 115–116
permitting, 116
treatment, 114–115
process description, 102–104
technical feasibility
description, 104–107

experience
composting case study, 111–112
land treatment case study, 108
oily sludges case study, 108–111
landfarming (*See* land treatment)
leaching and chemical reaction
economic feasibility
capital costs, 63
installation costs, 63
operation and maintenance costs, 63
qualitative ranking of cost, 63
environmental feasibility
environmental effectiveness, 62
exposure pathways, 62
general description, 52
implementation feasibility
design, 55–56
disposal, 60–61
equipment, 57–60
monitoring, 61
permitting, 61
treatment, 60
process description, 52–53
technical feasibility
description, 53
experience
bench and pilot-scale case studies, 53–55
low temperature thermal stripper (LTTS), 121–122, 128, 133
lower explosive limit (LEL), 189
lower flammable limit (LFL), 189

molten glass, 196
molten salt, 197

National Electrical Code (NEC), 31
National Fire Protection Association (NFPA), 31
National Pollution Discharge Elimination System (NPDES) permit, 61, 73, 97, 171, 181
no-action alternative, 79, 82
non-in situ technologies
asphalt incorporation (*See* asphalt incorporation)
chemical extraction (*See* chemical extraction)
excavation (*See* excavation)
future, 196–197
groundwater extraction and treatment (*See* groundwater extraction and treatment)
land treatment (*See* land treatment)
solidification/stabilization (*See* solidification/stabilization)
thermal treatment (*See* thermal treatment)

obsidian, 65

Pacific Northwest Laboratories (PNL)
electric melter technology, 64
in situ vitrification, 67
passive remediation (in situ)
economic feasibility
capital costs, 83–84
installation costs, 84
operation and maintenance costs, 84
qualitative ranking of cost, 84
environmental feasibility
environmental effectiveness, 83
exposure pathways, 83
general description, 75
implementation feasibility
design, 80–81
disposal, 82
equipment, 81
monitoring, 82
permitting, 82
treatment, 81
technical and process description, 75–79
technical feasibility
experience, 79–80
plasma system, 197
pozzolan (*See* solidification /stabilization)
publicly owned treatment works (POTW), 60–61

radio-frequency destruction, 195
recovery methods
forced, 56
gravity, 56
remedial technologies
in situ, 5 (*See also* in situ technologies)
non-in situ, 5 (*See also* non-in situ technologies)
state practices, 15
remediation (*See* passive remediation)

Research, Development and Demonstration (RD&D) permit, 132
Resource Conservation and Recovery Act (RCRA), 1, 131
risk assessment
  exposure assessment, 7–8
  hazard identification, 7–8
rotating kiln, 119–121
rototilling, 114

Seasonal Soil Compartment Model (SESOIL), 9–11
sheet piling, 92
slurry walls, 89
soil
  characteristics for land treatment, 112
  factors affecting biodegradation, 41–43
  factors affecting rate of volatilization, 20–22
  freezing, 197
  shredding, 197
solidification/stabilization
  economic feasibility
    capital costs, 155
    installation costs, 155
    operation and maintenance costs, 156
    qualitative ranking of cost, 156–157
  environmental feasibility
    environmental effectiveness, 154
    exposure pathways, 153–154
  general description, 143–144
  implementation feasibility
    design, 151
    disposal, 152
    equipment, 151–152
    monitoring, 152–153
    permitting, 153
    treatment, 152
  process description, 144
  technical feasibility
    cement-based processes, 148
    description, 145–148
    experience, 149–150
    pozzolanic or lime-based processes, 148–149
stabilization (*See* solidification/stabilization)
supercritical extraction, 197
supercritical leaching, 196
supercritical water oxidation, 197

temporary operating authorization (TOA), 132
thermal treatment
  economic feasibility
    capital costs, 133–134
    installation costs, 134
    operation and maintenance costs, 134–135
    qualitative ranking of cost, 135
  environmental feasibility
    environmental effectiveness, 133
    exposure pathways, 133
  general description, 118
  implementation feasibility
    design, 129
    disposal, 131
    equipment, 129–130
    monitoring, 131
    permitting, 131–132
    selection criteria, 128
    treatment, 130–131
  process description
    fluidized bed, 121
    low temperature thermal stripper (LTTS), 121–122
    rotating kiln, 119–121
  technical feasibility
    description, 124
    experience
      circulating bed incinerator case study, 126
      low temperature thermal stripper case study, 126–127
      rotary kiln case study, 126
trichloroethylene (TCE)
  in situ volatilization, 23–25

ultraviolet photolysis, 197
U.S. Department of Energy (DOE), 64
U.S. Environmental Protection Agency (EPA)
  extraction procedure, 153
underground storage tanks (UST)
  federal regulation, 1
upper explosive limit (UEL), 189
upper flammable limit (UFL), 189
Utility Solid Waste Activities Group (USWAG), 1

vitrification (in situ)
  economic feasibility
    capital costs, 74
    installation costs, 74
    operations and maintenance costs, 74–75
    qualitative ranking of cost, 75
  environmental feasibility
    environmental effectiveness, 73–74
    exposure pathways, 73
  general description, 64
  implementation feasibility
    design, 69
    disposal, 71
    equipment, 69
    monitoring, 71
    permitting, 73
    treatment, 71
  process description, 64–65
  technical feasibility
    description, 65–67
    experience
      engineering-scale case study, 67
volatilization (in situ)
  economic feasibility
    capital costs, 36
    installation costs, 36
    operation and maintenance costs, 37
    qualitative ranking of cost, 37
  environmental feasibility
    environmental effectiveness, 35–36
    exposure pathways, 35
  general description, 17
  implementation feasibility
    design, 31
    disposal, 33
    equipment, 31–32
    monitoring, 33–34
    permitting, 34
    treatment, 32–33
  process description, 17
  technical feasibility
    chemical factors, 23
    environmental factors, 22–23
    description, 18–23
    experience
      full-scale system, 25–26
      full-scale positive pressure system, 26
      pilot-scale system, 23–24
    management factors, 23
    soil factors, 20–22
    technical description, 18–23